Riesz and Fredholm theory in Banach algebras

B A Barnes, G J Murphy
M R F Smyth & T T West

University of Oregon/Dalhousie University/
Department of Health and Social Services, Northern Ireland/
Trinity College, Dublin

Riesz and Fredholm theory in Banach algebras

Pitman Advanced Publishing Program
BOSTON · LONDON · MELBOURNE

PITMAN BOOKS LIMITED
128 Long Acre, London WC2E 9AN

PITMAN PUBLISHING INC
1020 Plain Street, Marshfield, Massachusetts

Associated Companies
Pitman Publishing Pty Ltd, Melbourne
Pitman Publishing New Zealand Ltd, Wellington
Copp Clark Pitman, Toronto

First published 1982

AMS Subject Classifications: (main) 47B05, 47B30, 47B40
(subsidiary) 46BXX, 46JXX

British Library Cataloguing in Publication Data

Riesz and Fredholm theory in Banach algebras.—
(Research notes in mathematics; 67)
1. Banach algebras
I. Barnes, B. A. II. Series
512′.55 QA326

ISBN 0-273-08563-8

Library of Congress Cataloging in Publication Data
Main entry under title:

Riesz and Fredholm theory in Banach algebras.

(Research notes in mathematics; 67)
Bibliography: p.
Includes index.
1. Banach algebras. 2. Spectral theory
(Mathematics) I. Barnes, B. A. (Bruce A.)
II. Series
QA326.R54 512′.55 82-7550
ISBN 0-273-08563-8 AACR2

ISBN 0 273 08563 8

Reproduced and printed by photolithography
in Great Britain by Biddles Ltd, Guildford

Contents

Introduction

This monograph aims to highlight the interplay between algebra and spectral theory which emerges in any penetrating analysis of compact, Riesz and Fredholm operators on Banach spaces. The emphasis on algebra means that the setting within which most of the work takes place is a complex Banach algebra, though, in certain situations in which topology is dispensable, the setting is simply an algebra over the complex field. The choice of spectral theory as our second main theme means that there is little overlap with other extensions of classical results such as the study of Fredholm theory in von-Neumann algebras.

We use the monograph 'Calkin Algebras and Algebras of Operators in Banach Spaces' by Caradus, Pfaffenberger and Yood (25) as our take-off point, and it should be familiar, or at least accessible, to the reader. (A modern view of the Calkin algebra is given in (40)). The original intention behind Chapter O was to provide a summary of classical operator theory, but, it emerged in the course of the work that a quotient technique developed by Buoni, Harte and Wickstead (17), (41) led to new results, including a geometric characterisation of Riesz operators (§O.3) and some range inclusion theorems (§O.4). Thus Chapter O contains an amount of new material as well as a survey of classical results.

On an infinite dimensional Banach space a Fredholm operator is one which, by Atkinson's characterisation, is invertible modulo the ideal of finite rank operators (the socle of the algebra of all bounded linear operators on the Banach space). This motivates our concept of a Fredholm element in an algebra as one that is invertible modulo a particular ideal which, in the semisimple case, may be chosen to be the socle.

In §F.1 we introduce the left and right Barnes idempotents. For a Fredholm element in a semisimple algebra these always exist and lie in the socle. In the classical theory they are finite rank projections related to the kernel and range of a Fredholm operator. Primitive Banach algebras are considered in §F.2. Smyth has shown how the left regular representation of the algebra on a Banach space consisting of a minimal left ideal may be used to connect Fredholm elements in the algebra with Fredholm operators on the space. With this technique the main results of Fredholm theory in primitive algebras may be deduced directly from the classical results on Fredholm operators. This theory is extended in §F.3 to general Banach algebras by quotienting out the primitive ideals. It now becomes appropriate to introduce the index function (defined on the space of primitive ideals). The validity of both the index and punctured neighbourhood theorems in this general setting (first demonstrated by Smyth (83)) ensures that the full range of classical spectral theory of Fredholm (and of Riesz) operators carries over to Banach algebras.

Riesz theory is developed in Chapter R building on the Fredholm theory of the previous chapter and we follow Smyth's analysis (85) of the important class of Riesz algebras. Results which are peculiar to Hilbert space and their extensions to C*-algebras, including the West and Stampfli decomposition theorems are given in Chapter C*. Chapter A contains applications of our theory to seminormal elements in C*-algebras, operators leaving a fixed subspace invariant, triangular operators on sequence spaces, quasitriangular operators and measures on compact groups. The underlying algebraic requirements are listed in Chapter BA. Each chapter contains a final section of notes and comments.

A fair proportion of the theory developed here is appearing in print for the first time. Among the more important new results are the geometric characterisation of Riesz operators (O.3.5); the range inclusion theorems (§O.4); the link between Fredholm theory in primitive algebras and classical operator theory (F.2.6); the punctured neighbourhood theorem (F.2.10); the index function theorem (F.3.11); the characterisation of inessential ideals (R.2.6) and the Stampfli decomposition in C*-algebras (C*.2.6). (Some of these results have, however, been known since the appearance of (83)). This has required that full details of proofs be given, except for the theorems listed under the notes at the end of each chapter.

Each author has been involved in the development of the ideas presented in this monograph. The subject has gone through a period of rapid expansion and it now seems opportune to offer a unified account of its main results.

O Operator theory

This chapter contains the basic results from operator theory on Banach spaces often stated without proof. The main reference is the monograph of Caradus, Pfaffenberger and Yood (25). Bonsall (13) gives an algebraic approach to the spectral theory of compact operators; Schechter (80) and Heuser (43) are useful references for Fredholm theory; Dowson (29) and Heuser (44) are recommended for Riesz theory; while Dunford and Schwartz (30) provides an invaluable background of general spectral theory.

Notation and general information is set out in §1. Fredholm operators are considered in §2 which contains a proof of the Atkinson characterisation (O.2.2). §3 outlines the theory of Riesz operators and, employing a quotient technique due to Buoni, Harte and Wickstead (17), (41), contains a proof of the Ruston characterisation as well as a new geometric characterisation of Riesz operators due to Smyth (O.3.5). This material is used in §4 to prove range inclusion theorems for compact, quasinilpotent and Riesz operators several of which are new. Much simpler proofs of these results are available in Hilbert space and are given in §C*.5. In §5 we consider the action of a compact or Riesz operator on its commutant, and in §6 the properties of the wedge operator.

O.1 Notation

$\mathbb{R}$ and $\mathbb{C}$ will denote the real and complex fields, respectively, and X and H a Banach and a Hilbert space over $\mathbb{C}$. We start by listing the various classes of bounded linear operators which will be discussed and, where necessary, defined subsequently:

$\mathcal{B}(X)$ the Banach algebra of bounded linear operators on X;

$\mathrm{Inv}(\mathcal{B}(X))$ the set of invertible operators in $\mathcal{B}(X)$;

$\mathcal{F}(X)$ the ideal of finite rank operators on X;

$\mathcal{K}(X)$ the closed ideal of compact operators on X;

$\mathcal{I}(X)$ the closed ideal of inessential operators on X;

$\Phi(X)$ the set of Fredholm operators on X;

$\mathcal{Q}(X)$ the set of quasinilpotent operators on X;

$\mathcal{R}(X)$ the set of Riesz operators on X.

If $T \in \mathcal{B}(X)$, $\rho(T)$, $\sigma(T)$ and $r(T)$ will denote the resolvent set, spectrum and spectral radius of T, respectively. $\ker(T)$ will be the kernel of T and if Y is a subspace of X, invariant under T, $T|Y$ denotes the restriction of T to Y. X^* is the dual space of X and T^* the adjoint operator on X^*. If $x \in X$, $\alpha \in X^*$, $\alpha \otimes x$ is the operator of rank ≤ 1, $y \to \alpha(y)x$ on X. Let $\mathrm{Hol}(\sigma(T))$ denote the family of complex valued functions which are analytic in some neighbourhood of $\sigma(T)$. If $f \in \mathrm{Hol}(\sigma(T))$ the operator $f(T)$ is defined by the Cauchy integral

$$f(T) = \frac{1}{2\pi i}\int_\Gamma f(z)(z-T)^{-1}dz$$

where Γ is a suitable contour in $\rho(T)$ surrounding $\sigma(T)$. A subset ω of $\sigma(T)$ which is open and closed in $\sigma(T)$ is a *spectral set* for T. Associated with each spectral set ω is the *spectral projection* $P(\omega,T)$ defined by

$$P(\omega,T) = \frac{1}{2\pi i}\int_\Gamma f(z)(z-T)^{-1}dz$$

where Γ is a suitable contour in $\rho(T)$ surrounding $\sigma(T)$, and $f \in \mathrm{Hol}(\sigma(T))$ is one on ω and zero on $\sigma(T)\setminus\omega$. We use the following notation for projections. Let $P^2 = P \in \mathcal{B}(X)$ with $P : X = X_1 \oplus X_2$ where X_1 is the range and X_2 the kernel of P. If P reduces (commutes with) T we then write $P : T = T_1 \oplus T_2$, where $T_1 = T|X_1$, $T_2 = T|X_2$. If ω is a spectral set for T then $P(\omega,T) : T = T_1 \oplus T_2$ and $\sigma(T_1) = \omega$, $\sigma(T_2) = \sigma(T)\setminus\omega$. If λ is an isolated point of $\sigma(T)$ the corresponding spectral projection is written $P(\lambda,T)$. λ is called a *pole of finite rank* of T if $P(\lambda,T) \in \mathcal{F}(X)$. It is easy to check that $P(\lambda,T)$ is then the residue of the resolvent operator function $z \to (z-T)^{-1}$ at the point λ.

$\dim(Y)$ will denote the dimension of the space Y and U will denote the closed unit ball of X. If Y is a closed subspace of the Banach space X, X/Y denotes the quotient space of cosets $x + Y$; it is a Banach space under the norm $||x + Y|| = \inf_{y \in Y}||x + y||$. $\mathbb{Z}(\mathbb{Z}^+)$ denotes the set of integers (positive integers). The closure of a subset S of a topological space will be denoted $\bar{S}$.

The termination of a proof will be signified by ●

0.2 Fredholm operators

Let X be a Banach space over $\mathbb{C}$. $T \in \mathcal{B}(X)$ is of *finite rank* if $\dim(T(X)) < \infty$. T is a *compact* operator on X if $\overline{T(U)}$ is compact where U is the closed unit ball of X. The finite rank operators in $\mathcal{B}(X)$ form an ideal $\mathcal{F}(X)$ and the compact operators a closed ideal $\mathcal{K}(X)$. If $T \in \mathcal{K}(X)$ the Riesz theory of compact operators states that each non-zero point of $\sigma(T)$ is a pole of finite rank of T. $T \in \mathcal{B}(X)$ is a *Fredholm* operator if $\dim(\ker(T)) < \infty$, if $T(X)$ is closed in X, and if $\dim(X/T(X)) < \infty$. The set of Fredholm operators is denoted $\Phi(X)$. It follows from the Riesz theory that if $T \in \mathcal{K}(X)$ and $\lambda \neq 0$ then $\lambda - T \in \Phi(X)$.

The quotient algebra $\mathcal{B}(X)/\mathcal{K}(X)$ whose elements are the cosets $T + \mathcal{K}(X)$ is a Banach algebra under the quotient norm. It is called the *Calkin algebra* and will play a major role in our deliberations. Our immediate aim is to characterise Fredholm operators.

0.2.1 DEFINITION. (i) $\ell_\infty(X)$ is the linear space of bounded sequences $\{x_n\}$ of elements $x_n \in X$ with the supremum norm

$$||\{x_n\}|| = \sup_n ||x_n||.$$

(ii) $m(X)$ is the linear subspace of $\ell_\infty(X)$ consisting of those sequences every subsequence of which contains a convergent subsequence, i.e. those sequences with totally bounded sets of terms.

It is elementary to check that $\ell_\infty(X)$ is a Banach space and $m(X)$ a closed subspace of $\ell_\infty(X)$. Further, if $T \in \mathcal{B}(X)$ then $\{x_n\} \in \ell_\infty(X) \Rightarrow \{Tx_n\} \in \ell_\infty(X)$, and $\{x_n\} \in m(X) \Rightarrow \{Tx_n\} \in m(X)$. Let $\hat{X}$ denote the quotient space $\ell_\infty(X)/m(X)$, and if $T \in \mathcal{B}(X)$ let $\hat{T}$ denote the operator on $\hat{X}$ defined by

$$\hat{T}(\{x_n\} + m(X)) = \{Tx_n\} + m(X).$$

Clearly $\hat{T} \in \mathcal{B}(\hat{X})$, and $T \in \mathcal{K}(X) \Leftrightarrow \hat{T} = \hat{0}$.

0.2.2 THEOREM. (Atkinson characterisation) *For* $T \in \mathcal{B}(X)$ *the following statements are equivalent*

(i) $T \in \Phi(X)$;

(ii) $T + F(X) \in \mathrm{Inv}(\mathcal{B}(X)/F(X))$;

(iii) $T + K(X) \in \mathrm{Inv}(\mathcal{B}(X)/K(X))$;

(iv) $\hat{T} \in \mathrm{Inv}(\mathcal{B}(\hat{X}))$.

Proof. (i) => (ii). $T \in \Phi(X) \Rightarrow \dim(\ker(T)) < \infty$ and $T(X)$ is of finite co-dimension, so there exist closed subspaces Z and W of X such that

$$X = \ker(T) \oplus Z = T(X) \oplus W.$$

T can be depicted as the 2×2 operator matrix

$$T = \begin{array}{c} \\ W \\ T(X) \end{array} \begin{array}{cc} \ker(T) & Z \\ \hline \multicolumn{1}{|c|}{\phantom{T_{22}}} & \multicolumn{1}{c|}{\phantom{T_{22}}} \\ \hline \multicolumn{1}{|c|}{} & \multicolumn{1}{c|}{T_{22}} \\ \hline \end{array},$$

the subspaces on the top being domains and those on the left ranges; the unmarked entries are zero. $T_{22} : Z \to T(X)$ is bijective and continuous and $T(X)$ is closed so there exists a continuous linear inverse $S_{22} : T(X) \to Z$ ((30) p.57). If

$$S = \begin{array}{c} \\ \ker(T) \\ Z \end{array} \begin{array}{cc} W & T(X) \\ \hline \multicolumn{1}{|c|}{\phantom{S_{22}}} & \multicolumn{1}{c|}{\phantom{S_{22}}} \\ \hline \multicolumn{1}{|c|}{} & \multicolumn{1}{c|}{S_{22}} \\ \hline \end{array},$$

then

$$TS = \begin{array}{c} \\ W \\ T(X) \end{array} \begin{array}{cc} W & T(X) \\ \hline \multicolumn{1}{|c|}{} & \multicolumn{1}{c|}{} \\ \hline \multicolumn{1}{|c|}{} & \multicolumn{1}{c|}{I} \\ \hline \end{array}, \quad \text{and} \quad ST = \begin{array}{c} \\ \ker(T) \\ Z \end{array} \begin{array}{cc} \ker(T) & Z \\ \hline \multicolumn{1}{|c|}{} & \multicolumn{1}{c|}{} \\ \hline \multicolumn{1}{|c|}{} & \multicolumn{1}{c|}{I} \\ \hline \end{array}$$

Clearly, TS and ST are projections of finite co-dimension so there exist projections $P, Q \in F(X)$ such that

$$TS = I - P, \qquad ST = I - Q$$

so S is the inverse of T modulo $F(X)$.

(ii) ⟹ (iii) is obvious.

(iii) ⟹ (iv). If $S + K(X) = (T + K(X))^{-1}$, there exist $K_1, K_2 \in K(X)$ such that $TS = I - K_1$, $ST = I - K_2$. Clearly $\hat{S}\hat{T} = \hat{I} = \hat{T}\hat{S}$. (This argument is not reversible as we do not know that $\{\hat{T} : T \in B(X)\}$ is a closed subalgebra of $B(\hat{X})$).

(iv) ⟹ (i). Let $\hat{T} \in Inv(B(\hat{X}))$ and choose a sequence $\{x_n\}$ in the unit ball of ker(T). Then

$$\begin{aligned}\{Tx_n\} = 0 &\Rightarrow \hat{T}(\{x_n\} + m(X)) = 0,\\ &\Rightarrow \{x_n\} + m(X) = 0,\\ &\Rightarrow \{x_n\} \in m(X),\end{aligned}$$

so the unit ball of ker(T) is compact, hence $\dim(\ker(T)) < \infty$.

Next we show that $T(X)$ is closed in X. Since $\dim(\ker(T)) < \infty$, there exists a closed subspace Z of X such that $X = \ker(T) \oplus Z$. Clearly $T(X) = T(Z)$ and T is injective on Z so it is sufficient to prove that T is bounded below on Z. Suppose not; then there exists $\{x_n\} \subset Z$ with $||x_n|| = 1$ for each n and $Tx_n \to 0$.

$$\begin{aligned}\{Tx_n\} \in m(X) &\Rightarrow \hat{T}(\{x_n\} + m(X)) = 0,\\ &\Rightarrow \{x_n\} + m(X) = 0,\\ &\Rightarrow \{x_n\} \in m(X).\end{aligned}$$

Thus there exists a subsequence $\{x_{n_k}\}$ such that $x_{n_k} \to y \in X$. Then $||y|| = 1$ and $Tx_{n_k} \to Ty = 0$, but $Z \cap \ker(T) = (0)$ which is a contradiction.

Since $T(X)$ is closed, the quotient space $X/T(X)$ is a Banach space, it remains to prove $\dim(X/T(X)) < \infty$. Let $\{y_n\} \subset X$ be such that $||y_n + T(X)|| \leq 1$ for each n, then there exists $\{x_n\} \subset X$ such that $||y_n + Tx_n|| \leq 2$ for each n. $\hat{T}$ is invertible so there exists $\{w_n\} \in \ell_\infty(X)$ such that

$$\hat{T}(\{w_n\} + m(X)) = \{y_n + Tx_n\} + m(X),$$

thus $\{T(w_n - x_n) - y_n\} \in m(X)$,

hence there exists a subsequence such that

$$T(w_{n_k} - x_{n_k}) - y_{n_k} \to z \in X,$$

thus $||y_{n_k} + z + T(X)|| \to 0$ as $k \to \infty$

since $T(X)$ is closed. So $\{y_n + T(X)\}$ has a convergent subsequence, thus the unit ball of $X/T(X)$ is compact and $\dim(X/T(X)) < \infty$ ●

0.2.3 DEFINITION. If $T \in \mathcal{B}(X)$ the *essential spectrum* $\omega(T)$ is defined to be the spectrum of $T + \mathcal{K}(X)$ in the Calkin algebra.

0.2.4 COROLLARY. $\omega(T) = \sigma(\hat{T})$.

0.2.5 DEFINITION. (i) If $T \in \Phi(X)$ the *nullity* of T, $n(T) = \dim(\ker(T))$, the *defect* of T, $d(T) = \dim(X/T(X))$ and the *index* of T, $i(T) = i_X(T) = n(T) - d(T)$.

(ii) If $T \in \mathcal{B}(X)$, λ is a *Fredholm point* of T if $\lambda - T \in \Phi(X)$, and the set of Fredholm points of T is denoted by $\Phi(T)$.

Clearly $\rho(T) \subset \Phi(T) = \mathbb{C}\backslash\omega(T)$, further $\Phi(X)$ is a multiplicative semigroup and $i(T_1T_2) = i(T_1) + i(T_2)$ for $T_1, T_2 \in \Phi(X)$ ((25) 3.2.7). Moreover the set of Fredholm operators is invariant under compact perturbations and $i(T + K) = i(T)$ if $T \in \Phi(X)$, $K \in \mathcal{K}(X)$ ((25) 4.4.2). Perturbation theory is of crucial importance.

0.2.6 THEOREM. *If* $T \in \Phi(X)$ *there exists* $\delta > 0$ *such that* $S \in \mathcal{B}(X)$, $\|S\| < \delta \Rightarrow T + S \in \Phi(X)$ *and* $i(T + S) = i(T)$. ((25) 4.4.1).

Consequently, $\Phi(X)$ is an open semigroup in $\mathcal{B}(X)$ and the index function is continuous on $\Phi(X)$ and therefore constant on connected components of $\Phi(X)$. More detailed information is available if the perturbation is caused by a multiple of the identity.

0.2.7 THEOREM. *If* $T \in \Phi(X)$, *there exists* $\delta > 0$ *such that* $\lambda + T \in \Phi(X)$ *for* $|\lambda| \le \delta$ *and* $n(\lambda + T)$, $d(\lambda + T)$ *are constant and less than or equal to* $n(T)$, $d(T)$ *respectively for* $0 < |\lambda| < \delta$.

((25) 3.2.10). This important result we call the *punctured neighbourhood theorem*.

An index-zero Fredholm operator may be decomposed into the sum of an invertible operator plus a finite rank one.

0.2.8 THEOREM. $T \in \Phi(X)$ *and* $i(T) = 0 \Rightarrow$ *there exists* $F \in \mathcal{F}(X)$ *such that* $T + \lambda F \in \mathrm{Inv}(\mathcal{B}(X))$ $(\lambda \neq 0)$.

Proof. As in the proof of 0.2.2 we may write

$$T = \begin{array}{c|c|c|} & \ker(T) & Z \\ \hline W & & \\ \hline T(X) & & T_{22} \\ \hline \end{array}$$

where W and $\ker T$ have the same dimension since $i(T) = 0$. Construct $F \in \mathcal{F}(X)$ by means of the isomorphism $J : \ker(T) \to W$.

$$F = \begin{array}{c|c|c|} & \ker(T) & Z \\ \hline W & J & \\ \hline T(X) & & \\ \hline \end{array}$$

If $\lambda \neq 0$, $T + \lambda F \in \mathrm{Inv}(\mathcal{B}(X))$ ●

0.2.9 THEOREM. *If* $T \in \mathcal{B}(X)$ *then* $\sigma(T) \setminus \{\lambda \in \Phi(T)$ *and* $i(\lambda - T) = 0\}$

$$= \bigcap_{K \in \mathcal{K}(X)} \sigma(T + K).$$

Proof. The result may be restated as follows:

$$\{\lambda \in \Phi(T) \text{ and } i(\lambda - T) = 0\} = \bigcup_{K \in \mathcal{K}(X)} \rho(T + K).$$

Let $\lambda \in \bigcup_{K \in \mathcal{K}(X)} \rho(T + K)$; then for some $K_0 \in \mathcal{K}(X)$, $\lambda \in \rho(T + K_0)$ hence $\lambda - T - K_0 \in \Phi(X)$ and $i(\lambda - T - K_0) = 0$. But then $\lambda - T \in \Phi(X)$ and $i(\lambda - T) = 0$.

Conversely, let $\lambda - T \in \Phi(X)$ and $i(\lambda - T) = 0$. Without loss of generality take $\lambda = 0$. Then $T \in \Phi(X)$ and $i(T) = 0$ and, by 0.2.8, there exists $K_1 \in \mathcal{K}(X)$ such that $0 \in \rho(T + K_1)$ ●

0.3 Riesz operators

The ideal of *inessential* operators $\mathcal{I}(X)$ on a Banach space X is defined to be the inverse canonical image of the radical of the Calkin algebra. Since the radical is closed in the Calkin algebra, $\mathcal{I}(X)$ is closed in $\mathcal{B}(X)$. $T \in \mathcal{B}(X)$ is a *Riesz* operator if the non-zero spectrum of T consists of poles of finite rank of T.

If $x \in X$ and $\varepsilon > 0$, $\Delta(x,\varepsilon)$ denotes the open ball centred at x of radius ε.

0.3.1 DEFINITION. Let B be a bounded subset of X; the *measure of non-compactness* $q(B)$ of B is the infimum of $\varepsilon > 0$ such that B has a finite cover by open balls in X of radius ε.

If $B \subset \bigcup_1^n \Delta(x_i,\varepsilon)$ we say that $\{x_1, \dots x_n\}$ is a *finite* ε*-net* for B. (The points x_i need not necessarily lie in B). Thus

$$q(B) = \inf\{\varepsilon > 0 : \text{there exists a finite } \varepsilon\text{-net for } B\}.$$

Clearly $q(B) = 0 \iff B$ is totally bounded in X, so if B is closed in X, $q(B) = 0 \iff B$ is a compact subset of X.

0.3.2 LEMMA. *Let* B *be a bounded subset of* X, U *the closed unit ball of* X *and* $T \in \mathcal{B}(X)$. *Then*

$$q(T(U)) \le \sup_{q(B) \le 1} q(T(B)) \le 4q(T(U)).$$

Proof. The left hand inequality is obvious. To prove the right hand one suppose that $q(T(U)) < \varepsilon$ and let B be a bounded set such that $q(B) \leq 1$. Then

$$T(U) \subset \bigcup_{1}^{n} \Delta(y_i, \varepsilon), \qquad (y_1, \ldots y_n \in X)$$

$$\subset \bigcup_{1}^{n} \Delta(Tx_i, 2\varepsilon), \qquad (x_1, \ldots x_n \in U)$$

and $2T(U) \subset \bigcup_{1}^{n} \Delta(2Tx_i, 4\varepsilon)$.

Now $B \subset \bigcup_{1}^{m} \Delta(z_j, 1), \qquad (z_1, \ldots z_m \in X)$

$$\subset \bigcup_{1}^{m} \Delta(b_j, 2), \qquad (b_1, \ldots b_m \in B)$$

$$\subset \bigcup_{1}^{m} (b_j + 2U).$$

Thus $T(B) \subset \bigcup_{1}^{m} (Tb_j + 2T(U))$,

$$\subset \bigcup_{1}^{m} (Tb_j + \bigcup_{1}^{n} \Delta(2Tx_i, 4\varepsilon)),$$

$$\subset \bigcup_{1}^{m} \bigcup_{1}^{n} (Tb_j + \Delta(2Tx_i, 4\varepsilon)),$$

$$\subset \bigcup_{1}^{m} \bigcup_{1}^{n} \Delta(Tb_j + 2Tx_i, 4\varepsilon),$$

so $q(T(B)) \leq 4\varepsilon$ ●

0.3.3 LEMMA. *If* $\{x_n\} \in \ell_\infty(X)$, $q(\{x_n\}) = ||\{x_n\} + m(X)||$.

Proof. Let $||x_n + m(X)|| < \delta$, then there exists $\{y_n\} \in m(X)$ such that $||\{x_n\} - \{y_n\}|| < \delta$. Since $\{y_n\} \in m(X)$, there exists a finite ε-net for $\{y_n\}$ for each $\varepsilon > 0$. This is a finite $(\varepsilon+\delta)$-net for $\{x_n\}$. Thus $q(\{x_n\}) \leq \varepsilon + \delta$ and, as ε is arbitrary, $q(\{x_n\}) \leq \delta$. It follows that $q(\{x_n\}) \leq ||\{x_n\} + m(X)||$.

Conversely, let $q(\{x_n\}) < \delta$; then there exists a finite δ-net for $\{x_n\}$, say $y_1, \dots y_\ell$, so for each n there exists j $(1 \le j \le \ell)$ such that $||x_n - y_j|| < \delta$. If $z_n = y_j$ for each such n we obtain a sequence $\{z_n\} \varepsilon\ m(X)$. Now

$$||\{x_n\} - \{z_n\}|| < \delta,$$

hence $||\{x_n\} + m(X)|| < \delta$,

so $\quad ||\{x_n\} + m(X)|| \le q(\{x_n\})$ ●

Recall that $T \varepsilon \mathcal{B}(X)$ induces an operator $\hat{T} \varepsilon \mathcal{B}(\hat{X})$ where $\hat{X} = \ell_\infty(X)/m(X)$, by virtue of the equation

$$\hat{T}(\{x_n\} + m(X)) = \{Tx_n\} + m(X).$$

0.3.4 LEMMA. $||\hat{T}|| \le \sup\limits_{q(B)\le 1} q(T(B)) \le 2||\hat{T}||$.

<u>Proof</u>. $$\begin{aligned} ||\hat{T}|| &= \sup\{||\{Tx_n\} + m(X)|| : ||\{x_n\} + m(X)|| \le 1\}, \\ &= \sup\{q(\{Tx_n\}) : q(\{x_n\}) \le 1\}, \qquad (0.3.3) \\ &\le \sup_{q(B)\le 1} q(T(B)). \end{aligned}$$

Let B be a bounded subset of X such that $q(B) > \delta > 0$. Then we may choose an infinite sequence $\{x_n\}$ in B inductively so that $x_{n+1} \varepsilon B \setminus \bigcup_1^n \Delta(x_j, \delta)$ for $n \ge 1$. Clearly $||x_m - x_n|| > \delta$ for $m \ne n$ and so $q(\{x_n\}) > \delta/2$. If $\varepsilon > 0$ apply this to the set T(B) where $q(B) \le 1$, to obtain a sequence $\{Tx_n\}$ such that $q(\{Tx_n\}) > \frac{1}{2} q(T(B)) - \varepsilon$. Now

$$||\hat{T}|| = \sup\{q(\{Tx_n\}) : q(\{x_n\}) \le 1\} > \tfrac{1}{2} q(T(B)) - \varepsilon,$$

thus $\quad 2||\hat{T}|| \ge \sup\limits_{q(B)\le 1} q(T(B))$ ●

We have now, somewhat laborously, set up the machinery required for our characterisations of Riesz operators.

0.3.5 THEOREM. (Ruston characterisation) *For* $T \in \mathcal{B}(X)$ *the following statements are equivalent*

(i) $T \in \mathcal{R}(X)$;

(ii) $r(T + \mathcal{K}(X)) = 0$;

(iii) $r(\hat{T}) = 0$;

(iv) $\lim_n q(T^n(U))^{1/n} = 0$;

(v) for each $\varepsilon > 0$ there exists $n \in \mathbb{Z}^+$ such that $T^n(U)$ has a finite ε^n-net.

<u>Proof</u>. (i) <=> (ii). Let T be a Riesz operator, then if $0 \neq \lambda \in \sigma(T)$ the corresponding spectral projection $P(\lambda,T) \in \mathcal{F}(X)$. If $\delta > 0$ the set $\{\lambda \in \sigma(T) : |\lambda| > \delta\}$ is finite and the corresponding spectral projection $P \in \mathcal{F}(X)$. Now $r(T - TP) \leq \delta$ and $TP \in \mathcal{F}(X)$; then

$$r(T + \mathcal{K}(X)) \leq \inf_{K \in \mathcal{K}(X)} r(T + K) \leq \delta,$$

and since δ is arbitrary $r(T + \mathcal{K}(X)) = 0$.

Conversely, let $T \in \mathcal{B}(X)$ satisfy $\sigma(T + \mathcal{K}(X)) = \{0\}$. If $0 \neq \lambda$ then $\lambda - T + \mathcal{K}(X) \in \mathrm{Inv}(\mathcal{B}(X)/\mathcal{K}(X))$, so, by the Atkinson characterisation (0.2.2), $\lambda - T \in \Phi(X)$. If $0 \neq \lambda \in \partial\sigma(T)$ each neighbourhood of λ must contain points of $\rho(T)$, thus using the punctured neighbourhood theorem (0.2.7), $n(\lambda - T) = 0 = d(\lambda - T)$ for $0 < |\lambda| < \delta$, and some positive δ; therefore this punctured neighbourhood lies in $\rho(T)$ and λ is an isolated point of $\sigma(T)$. But if the non-zero boundary points of $\sigma(T)$ are all isolated, all non-zero points in $\sigma(T)$ must be isolated. Let λ be one such and Γ a contour in $\rho(T)$ surrounding λ but no other point of $\sigma(T)$. Then

$$P(\lambda,T + \mathcal{K}(X)) = P(\lambda,T) + \mathcal{K}(X) = \frac{1}{2\pi i}\int_\Gamma (z - T - \mathcal{K}(X))^{-1} dz = 0,$$

since $z - T - \mathcal{K}(X)$ is invertible inside and on Γ. So $P(\lambda,T) \in \mathcal{K}(X)$ and λ is a pole of finite rank of T.

(ii) $\Leftrightarrow$ (iii). This follows at once from 0.2.4.

(iii) $\Leftrightarrow$ (iv). Combining 0.3.2 and 0.3.4 we get

$$||\hat{T}|| \leq 4q(T(U)) \leq 8||\hat{T}||, \qquad (\S)$$

and the equivalence follows by considering $||\hat{T}^n||^{1/n}$.

(iv) $\Leftrightarrow$ (v). This is now clear since $T^n(U)$ has a finite ε^n-net $\Leftrightarrow$ $q(T^n(U)) < \varepsilon^n$ ●

An easy consequence of the Ruston characterisation and of properties of the spectral radius in the Calkin algebra is the following result. $[S,T] = ST - TS$ is the *commutator* of S and T.

0.3.6 THEOREM. (i) $S, T \in R(X)$ *and* $[S,T] \in K(X) \Rightarrow S + T \in R(X)$;

(ii) $S \in B(X)$, $T \in R(X)$ *and* $[S,T] \in K(X) \Rightarrow ST, TS \in R(X)$;

(iii) $T_n \in R(X)$ $(n \geq 1)$, $T \in B(X)$, $||T_n - T|| \to 0$ and $[T_n,T] \in K(X)$ $(n \geq 1) \Rightarrow T \in R(X)$.

Another useful consequence involves functions of a Riesz operator. Let $T \in B(X)$ and $f \in Hol(\sigma(T))$.

0.3.7 THEOREM. (i) $T \in R(X)$ *and* $f(0) = 0 \Rightarrow f(T) \in R(X)$;

(ii) *If* $T \in B(X)$ *and* $f(z)$ *does not vanish on* $\sigma(T)\setminus\{0\}$ *then* $f(T) \in R(X) \Rightarrow T \in R(X)$.

In fact $f(0) = 0 \Rightarrow f(T) = Tg(T)$, where $g \in Hol(\sigma(T))$ and $[T,g(T)] = 0$, hence 0.3.7(i) follows from 0.3.6(ii).

0.4 Range inclusion

The machinery developed in §3 allows us to deduce properties of an operator S from an operator T provided that $S(X) \subset T(X)$. In this section we shall use $S^{-1}(U)$ to denote the inverse image of U under S whether, or not, S is invertible.

0.4.1 THEOREM. *If* $S,T \in B(X)$ *and* $S(X) \subset T(X)$ *there exists* $\eta > 0$ *such that* $S(U) \subset \eta(\overline{T(U)})$.

Proof. $X = S^{-1}(T(X)) = S^{-1}(T(\bigcup_{n=1}^{\infty} n\, U)) \subset S^{-1}(\bigcup_{n=1}^{\infty} n\, \overline{T(U)}) = \bigcup_{n=1}^{\infty} n\, S^{-1}(\overline{T(U)}) = X$.

S is continuous hence $S^{-1}(\overline{T(U)})$ is closed in X therefore by the Baire category theorem ((30) p.20), $nS^{-1}\overline{(T(U))}$ has a non-empty interior for some $n \in \mathbb{Z}^+$. But $nS^{-1}\overline{(T(U))}$ is homeomorphic to $S^{-1}\overline{(T(U))}$, so there exist $x \in X$ and $\varepsilon > 0$ such that $S(\Delta(x,\varepsilon)) \subset \overline{T(U)}$. Hence there exists $\{x_n\} \subset U$ such that $Sx = \lim_n Tx_n$, and if $||y|| < \varepsilon$, there exists $\{z_n\} \subset U$ such that $S(x + y) = \lim_n Tz_n$. Thus $Sy = \lim_n T(z_n - x_n)$ and $\{z_n - x_n\} \subset 2U$. Finally, if $||y|| \leq 1$, there exists $\{y_n\} \subset U$ such that $Sy = \eta \lim_n Ty_n$ where $\eta = 2\varepsilon^{-1}$ ●

0.4.2 COROLLARY. $S(X) \subset T(X)$ *and* $T \in \mathcal{K}(X) \Rightarrow S \in \mathcal{K}(X)$.

0.4.3 THEOREM. *If* $S,T \in \mathcal{B}(X)$, $ST = TS$, *and* $S(U) \subset \eta\overline{(T(U))}$ *then* $S^n(U) \subset \eta^n\overline{(T^n(U))}$.

<u>Proof</u>. By hypothesis the result is true for $n = 1$. Suppose it is true for n, and let $y \in U$, $\delta > 0$. Then there exists $z \in U$ such that

$$||S^n y - \eta^n T^n z|| < \tfrac{1}{2}\delta||S||^{-1}.$$

Thus $||S^{n+1}y - \eta^n ST^n z|| < \frac{1}{2}\delta$,

$$\text{so} \quad ||S^{n+1}y - \eta^n T^n Sz|| < \tfrac{1}{2}\delta. \tag{†}$$

But there exists $w \in U$ such that

$$||Sz - \eta Tw|| < \tfrac{1}{2}\delta\eta^{-n}||T^n||^{-1},$$

$$\text{so} \quad ||\eta^n T^n Sz - \eta^{n+1}T^{n+1}w|| < \tfrac{1}{2}\delta. \tag{*}$$

From (†) and (*) we see that if $y \in U$, there exists $w \in U$ such that

$$||S^{n+1}y - \eta^{n+1}T^{n+1}w|| < \delta,$$

that is $S^{n+1}(U) \subset \eta^{n+1}\overline{T^{n+1}(U)}$ and the proof follows by induction ●

Combining 0.4.1 and 0.4.3 we get

0.4.4 COROLLARY. *If* $S(X) \subset T(X)$ *and* $ST = TS$ *then* $T \in \mathcal{Q}(X) \Rightarrow S \in \mathcal{Q}(X)$.

0.4.5 THEOREM. *If* $S, T \in \mathcal{B}(X)$, $S(X) \subset T(X)$ *and* $[S,T] \in K(X)$ *then* $T \in \mathcal{R}(X) \Rightarrow S \in \mathcal{R}(X)$.

<u>Proof</u>. Let V denote the closed unit ball of $\hat{X} = \ell_\infty(X)/m(X)$. If $\varepsilon > 0$ and $\{x_n\} + m(X) \in V$, there exists $\{y_n\} \in m(X)$ such that

$$||x_n + y_n|| < 1 + \varepsilon \qquad (n \geq 1),$$

then $\left\{ \dfrac{x_n + y_n}{1 + \varepsilon} \right\} \subset U$.

Now there exists $\eta > 0$ such that $S(U) \subset \overline{\eta T(U)}$ (0.4.1) so there exists $\{z_n\} \subset U$ such that

$$\left|\left| \frac{S(x_n + y_n)}{1 + \varepsilon} - \eta T z_n \right|\right| < \varepsilon \qquad (n \geq 1),$$

so $\quad ||S(x_n + y_n) - (1 + \varepsilon)\eta T z_n|| < \varepsilon(1 + \varepsilon) \qquad (n \geq 1)$,

hence $||S(x_n + y_n) - \eta T z_n|| < \varepsilon(1 + \varepsilon) + \varepsilon\eta||T|| \qquad (n \geq 1)$,

since $\{z_n\} \subset U$. Now $\{y_n\} \in m(X)$, hence $\{Sy_n\} \in m(X)$, therefore

$$||\hat{S}(\{x_n\} + m(X)) - \eta\hat{T}(\{z_n\} + m(X))|| < \varepsilon(1 + \varepsilon) + \varepsilon\eta||T||,$$

and since $\{x_n\} + m(X)$, $\{z_n\} + m(X) \in V$ we get

$$\hat{S}(V) \subset \overline{\eta\hat{T}(V)} \quad \text{and} \quad [S,T] \in K(X) \Rightarrow [\hat{S},\hat{T}] = 0, \quad \text{so, by 0.4.3,}$$

$$\hat{S}^n(V) \subset \overline{\eta^n(\hat{T}^n(V))} \qquad (n \geq 1),$$

which gives $\quad ||\hat{S}^n|| \leq \eta^n||\hat{T}^n||$,

thus $r(\hat{S}) \leq r(\hat{T})$. But $r(\hat{T}) = 0$ by 0.3.5, hence $r(\hat{S}) = 0$ and $S \in \mathcal{R}(X)$ again by 0.3.5 ●

0.5 Action on the commutant

If $T \in \mathcal{B}(X)$, $Z(T)$ denotes the commutant of T which is a closed subalgebra of $\mathcal{B}(X)$ and $\tau : S \to ST$ is the operator of multiplication by T on $Z(T)$. Obviously $||\tau|| = ||T||$. If τ is a compact (Riesz) operator on $Z(T)$ we say that T has a *compact (Riesz) action on its commutant.*

0.5.1 LEMMA. $\sigma(\tau) = \sigma(T)$.

Proof. $\lambda \in \rho(T) \Rightarrow (\lambda - T)^{-1} \in Z(T)$, and then

$$(\lambda - \tau)^{-1} : S \to S(\lambda - T)^{-1} \in \mathcal{B}(Z(T)),$$

hence $\lambda \in \rho(\tau)$.

Conversely, if $\lambda \in \rho(T)$, there exists $\nu \in \mathcal{B}(Z(T))$ such that $(\lambda - \tau)\nu = \nu(\lambda - \tau)$ is the identity on $Z(T)$. Thus

$$I = (\lambda - \tau)\nu(I) = (\lambda - T)\nu(I) = \nu(I)(\lambda - T)$$

since everything commutes, thus $\lambda \in \rho(T)$ ●

The next result states that if T is a compact operator on X then τ is also compact. As we remark (p. 20) the converse statement is false.

0.5.2 THEOREM. $T \in K(X) \Rightarrow \tau \in K(Z(T))$.

Proof. If $S_n \in Z(T)$, $||S_n|| = 1$ $(n \geq 1)$, we need to show that $\{\tau(S_n)\}_1^\infty = \{S_n T\}_1^\infty$ has a norm convergent subsequence. Let U be the closed unit ball of X and put $E = \overline{T(U)}$. If $S \in Z(T)$ and $||S|| \leq 1$, $ST(U) = TS(U) \subset T(U)$ hence by continuity $S(E) \subset E$. Now the restriction $S|E$ of S to E is contained in $C_X(E)$; the set of continuous functions mapping the compact Hausdorff space E to X. Since

$$||S|| \leq 1 \Rightarrow ||Sx - Sx'|| \leq ||x - x'||,$$

the set $\{S|E : S \varepsilon Z(T), ||S|| \leq 1\}$ is an equicontinuous subset of $\mathcal{C}_\chi(E)$ and is therefore, by the Arzelà-Ascoli theorem ((30) p.266), a compact subset of $\mathcal{C}_\chi(E)$. Hence if $S_n \varepsilon Z(T)$ and $||S_n|| \leq 1$ there exists a subsequence $\{S_{n_k}\}$ which converges uniformly on E, i.e. $\{S_{n_k}Tx\}$ converges uniformly on U, i.e. $\{S_{n_k}T\}$ is norm convergent ●

A more complete result is true for Riesz operators.

0.5.3 THEOREM. $T \varepsilon \mathcal{R}(X) \iff \tau \varepsilon \mathcal{R}(Z(T))$.

Proof. Let $\lambda \neq 0$. Lemma 0.5.1 shows that λ is an isolated point of $\sigma(T) \iff \lambda$ is an isolated point of $\sigma(\tau)$, and in this case the associated spectral projections are connected by the formula

$$P(\lambda,\tau) : S \to SP(\lambda,T) \qquad (S \varepsilon Z(T)).$$

Now if $S \varepsilon Z(T)$, then S commutes with $P(\lambda,T)$ so

$$P(\lambda,T) : X = X_1 \oplus X_2,$$

$$T = T_1 \oplus T_2,$$

$$S = S_1 \oplus S_2, \qquad (S \varepsilon Z(T))$$

$$P(\lambda,T)S = S_1 \oplus O_2, \qquad (S \varepsilon Z(T))$$

where $S_1 \varepsilon Z(T_1)$, therefore

$$P(\lambda,\tau)(Z(T)) = \{P(\lambda,T)S : S \varepsilon Z(T)\} = \{S_1 \oplus O_2 : S_1 \varepsilon Z(T_1)\}.$$

Suppose now that $T \varepsilon \mathcal{R}(X)$, then $0 \neq \lambda \varepsilon \sigma(T)$ is an isolated point of $\sigma(T)$ and $\dim(X_1) < \infty$. It follows that $\dim(Z(T_1)) \leq \dim(\mathcal{B}(X_1)) < \infty$, thus $\dim(P(\lambda,\tau)Z(T)) < \infty$ hence $\tau \varepsilon \mathcal{R}(Z(T))$.

Conversely, let $\tau \varepsilon \mathcal{R}(Z(T))$ and $0 \neq \lambda \varepsilon \sigma(\tau)$; then λ is an isolated point of $\sigma(\tau)$ and $\dim(P(\lambda,\tau)Z(T)) = \dim(Z(T_1)) < \infty$. Since the algebra generated by T_1 is contained in $Z(T_1)$ it must also be finite dimensional, thus there exists a non-zero polynomial p such that $p(T_1) = 0$. But $\sigma(T_1) = \{\lambda\}$ thus $(\lambda - T_1)^k = 0$ for some positive integer k. If $(\lambda - T_1)^{-1}(0)$ is infinite dimensional it contains an infinite linearly independent set $\{x_n\}$ and $T_1 x_n = \lambda x_n$ (for each n); also there exists $(0 \neq)\alpha \in X_1^*$ such that $T^*\alpha = \lambda\alpha$. It follows at once that the infinite linearly independent set $\{\alpha \otimes x_n\}$ of rank one operators lies in $Z(T_1)$. Thus $(\lambda - T_1)^{-1}(0)$ is finite dimensional, hence so is $(\lambda - T_1)^{-k}(0) = X_1$, thus $T \in \mathcal{R}(X)$ ●

0.6 The wedge operator

If $T \in \mathcal{B}(X)$ we define the *wedge operator* $T \wedge T$ on $\mathcal{B}(X)$ by

$$T \wedge T : S \to TST \quad (S \in \mathcal{B}(X)).$$

Clearly $||T \wedge T|| = ||T||^2$.

0.6.1 THEOREM. $T \in \mathcal{B}(X)$ *is (i) finite rank, (ii) compact, (iii) Riesz, or (iv) quasinilpotent* $\Leftrightarrow$ $T \wedge T$ *is.*

Proof. (i) Let $T = \sum_{i=1}^{n} \alpha_i \otimes x_i$, then

$$TST = \left(\sum_{i=1}^{n} \alpha_i \otimes x_i\right) S \left(\sum_{j=1}^{n} \alpha_j \otimes x_j\right) = \sum_{i,j=1}^{n} \alpha_i(Sx_j)\alpha_j \otimes x_i.$$

Thus $\{TST : S \in \mathcal{B}(X)\} \subset \text{span}\,\{\alpha_j \otimes x_i : 1 \leq i, j \leq n\}$

which is a finite dimensional subspace of $\mathcal{B}(X)$.

Conversely, suppose that $T \neq 0$ and $T \wedge T$ is finite rank. If $T(X)$ contains an infinite linearly independent set $\{Tx_i\}_1^\infty$ choose $\alpha \in X^*$ such that $T^*\alpha \neq 0$ and then the set

$$\{T \wedge T(\alpha \otimes x_i)\}_1^\infty = \{T(\alpha \otimes x_i)T\}_1^\infty = \{T^*\alpha \otimes Tx_i\}_1^\infty$$

is an infinite linearly independent set in $T \wedge T(\mathcal{B}(X))$.

(ii) Let U be the closed unit ball of X and $\mathcal{B}_1$ the closed unit ball of $\mathcal{B}(X)$. If T is compact, $T(U)$ is totally bounded, hence, if $\varepsilon > 0$ there exist $x_1, \ldots x_k \in U$ such that for each $x \in U$ there exists i $(1 \leq i \leq k)$ with $||T(x - x_i)|| < \varepsilon$. Fix i, then the set $\{STx_i : S \in \mathcal{B}_1\}$ is bounded in X, so the set $\{TSTx_i : S \in \mathcal{B}_1\}$ is totally bounded. Thus the set $\{(TSTx_1, \ldots, TSTx_k) : S \in \mathcal{B}_1\}$, which is contained in the product space $X^{(k)}$ in the product topology, is a subset of a totally bounded set, hence is itself totally bounded. Therefore there exist $S_1, \ldots S_n \in \mathcal{B}_1$ such that each point of the above set is within ε of $(TS_jTx_1, \ldots, TS_jTx_k)$ for some j $(1 \leq j \leq n)$. If $x \in U$ and $S \in \mathcal{B}_1$

$$||TSTx - TS_jTx|| \leq ||T|| \; ||S|| \; ||T(x - x_i)|| + ||TSTx_i - TS_jTx_i||$$

$$+ ||T|| \; ||S_j|| \; ||Tx_i - Tx|| \leq (2||T|| + 1)\varepsilon.$$

Thus $||TST - TS_jT|| \leq (2||T|| + 1)\varepsilon$ and the set $\{TAT : A \in \mathcal{B}_1\}$ is totally bounded, i.e. $T \wedge T$ is compact.

Conversely, let $T \neq 0$ and $T \wedge T$ be compact. If $\{x_i\}$ is a bounded sequence in X and $0 \neq \alpha \in X^*$, $\{\alpha \otimes x_i\}$ is a bounded sequence in $\mathcal{B}(X)$, thus

$$\{T \wedge T(\alpha \otimes x_i)\} = \{T^*\alpha \otimes Tx_i\}$$

is a bounded sequence in $\mathcal{B}(X)$ and thus has a subsequence such that

$$||T^*\alpha \otimes Tx_{i_n} - T^*\alpha \otimes Tx_{i_m}|| = ||T^*\alpha|| \; ||Tx_{i_n} - Tx_{i_m}|| \to 0.$$

Hence $\{Tx_i\}$ has a convergent subsequence and T is compact.

(iii) If T is a Riesz operator, then for $\varepsilon > 0$ there exists a compact K_n such that $||T^n - K_n|| < \varepsilon^n$ $(n \geq N)$. Without loss of generality take $||T||, ||K_n|| < 1$.

$$||(T \wedge T)^n - K_n \wedge K_n|| = \sup_{||S|| \leq 1} ||T^nST^n - K_nSK_n|| \leq 2||T^n - K_n|| < 2\varepsilon^n.$$

$K_n \wedge K_n$ is compact by part (ii), so, using the Ruston characterisation (0.3.5), $T \wedge T$ is Riesz.

Conversely, let $T \wedge T$ be Riesz on $\mathcal{B}(X)$; then its restriction $T \wedge T|Z(T)$ is a Riesz operator ((25) 3.5.1), i.e. $S \to ST^2$ is Riesz on $Z(T)$ and by Theorem 0.5.3, T^2 is Riesz on X, hence, by the Ruston characterisation, so is T.

(iv) $\quad ||(T \wedge T)^n|| = ||T^n \wedge T^n|| = ||T^n||^2.$

Thus $\quad r(T \wedge T) = \lim_n ||(T \wedge T)^n||^{1/n} = \lim_n ||T^n||^{2/n} = r(T)^2.$

Clearly $T \wedge T$ is quasinilpotent $\Leftrightarrow$ T is ●

0.7 Notes

Compact or completely continuous operators arose first in the guise of quadratic forms in Hilbert's work (46) in 1906. Fredholm, three years earlier, had considered resolvent expansions of certain integral operators (33). The spectral theory of compact operators was worked out by F. Riesz (74) in 1918 for the important special case in which the underlying Banach space is a space of continuous functions. Fredholm theory in its present form dates from the work of Atkinson (5) and Gohberg (36), (37) in 1951 and Yood (102), 1954. The characterisation of Fredholm operators as those which are invertible modulo the compact or finite rank operators (0.2.2) is due to Atkinson. Ruston started work on Riesz operators (76) in 1954 while Calkin's work on ideals of operators in Hilbert space (20) had appeared in 1941 and Kleinecke (53) introduced the important ideal of inessential operators, which is the biggest closed ideal contained in the Riesz operators, in 1963. The properties of Riesz operators were examined by Heuser (42) in 1963; in more detail by Caradus (21) and West (94), (95) in 1966. This class is of interest on account of the spectral properties which Riesz operators share with compact operators. From the algebraic standpoint, if commutativity (or commutativity modulo the compact operators) is added to a result valid for $\mathcal{K}(X)$, it often extends to $\mathcal{R}(X)$. Examples of this phenomenon occur in 0.3.6 and 0.4.5.

The technique, set up in §0.2, and used to study Riesz operators in §0.3 is based on work of Buoni, Harte and Wickstead (17), (41). Observe that by replacing T by T^n in inequality (§) on p.12, then taking n-th roots and using 0.2.4 we get

$$r(\hat{T}) = r(T + K(X)) = \lim_n q(T^n(U))^{1/n} .$$

Further information on measures of non-compactness and the essential spectrum is given by Zemánek (107), (108). The representation $T + K(X) \to \hat{T}$ of the Calkin algebra on X as a subalgebra of $B(\hat{X})$ is due to Lebow and Schechter (57). It is not known if this representation is irreducible, or if its image is closed in $B(\hat{X})$, although Lebow and Schechter have shown that the latter statement is true if X satisfies the Grothiendieck approximation property ((57) 3.7). The equivalence of (i) and (iv) in 0.2.2 is due to Buoni, Harte and Wickstead ((17) Theorem 3). If T is an index-zero Fredholm operator on a Banach space Murphy and West (62) followed by Laffey and West (55) have strengthened 0.2.8 by showing that $T = V + F$ where V is invertible, F is of finite rank and $[V,F]^2 = 0$. Note that $[V,F] = 0 \iff$ zero is either in the resolvent set of T, or is an isolated pole of T of finite rank ((25) 1.4.5), and in this case T is called a *Riesz-Schauder* operator. Ruston proved the equivalence of (i) and (ii) in 0.3.5. Condition (iv) is due to G.J. Murphy and the geometric characterisation (v) of Riesz operators to M.R.F. Smyth.

The range inclusion theorem for compact operators (0.4.2) is attributed to R.S. Phillips; the result for quasinilpotent operators (0.4.4) is due to M.J. Ganly, while the result for Riesz operators (0.4.5) is due to M.R.F. Smyth.

The fact that a compact operator acts compactly on its commutant (0.5.2) was observed by Bonsall (11), (12) and used to develop an algebraic approach to the spectral theory of compact operators. Theorem 0.5.3 showing that a Riesz operator has a Riesz action on its commutant and its converse is due to Smyth (81). The converse of 0.5.2 was shown to be false by I.S. Murphy (63), with an ingenious counter-example of a weighted shift operator on ℓ_1. Murphy overcame the main difficulty, which is to determine the commutant, by choosing the weights so that it is the strongly closed algebra generated by the operator.

The wedge operator $T \wedge T$ was introduced by Vala (91) and its properties were studied by Alexander (4), who also considered the generalised wedge operator defined by

$$S \wedge T : V \to SVT \quad (V \in B(X)).$$

The following result generalises 0.6.1.

0.7.1 THEOREM. *If* $S,T \in \mathcal{B}(X)$ *then* S *and* T *belong to classes* (i), (ii) *or* (iii) $\iff$ $S \wedge T$ *does*

(i) *the non-zero finite rank operators;*

(ii) *the non-zero compact operators;*

(iii) *the Riesz operators which are not quasi-nilpotent. Further* $S \wedge T$ *is quasinilpotent* $\iff$ *either* S *or* T *is quasinilpotent.*

F Fredholm theory

In O.2.2 we saw that a bounded linear operator on an infinite dimensional Banach space X is Fredholm $\Leftrightarrow$ it is invertible modulo $F(X) \Leftrightarrow$ it is invertible modulo $K(X)$. The fact that $B(X)$ is a primitive algebra ((O) is a primitive ideal) and that $F(X)$ is the socle of $B(X)$ motivates the work of this chapter.

Fredholm theory in Banach algebras was pioneered by Barnes (7), (8) in semisimple and semiprime algebras. This theory was extended to general Banach algebras by Smyth (83). Here we adopt a simpler approach due again to Smyth and use the case of a primitive Banach algebra, where Fredholm theory is straightforward, with natural analogues of the rank nullity, defect and index of a Fredholm operator for Fredholm elements of the algebra (those invertible modulo the socle), to build up to the general case.

§1 contains information on minimal ideals and minimal idempotents, also in F.1.10 we exhibit the Barnes idempotents for a Fredholm element. These are fundamental to the rest of the chapter. Fredholm theory in a primitive Banach algebra is developed in §2; the nullity, defect and index of a Fredholm element in a primitive algebra are defined, and a representation connects these quantities with the nullity, defect and index of a certain Fredholm operator. This allows us to deduce the important results of Fredholm theory in primitive Banach algebras directly from their counterparts in operator theory. If P is a primitive ideal of a general Banach algebra A then the quotient algebra A/P is primitive. This fact is exploited in §3 to obtain a Fredholm theory in the general case. As observed by Pearlman (69) and Smyth (83) it is necessary to replace the numerical valued nullity, defect and index of the primitive case by functions (of finite support) defined (for each Fredholm element) on the structure space of the algebra. §4 besides containing observations on extensions of the theory also contains results on generalised indices and on the algebraic kernel (the largest ideal of algebraic elements).

F.1 Minimal ideals and Barnes idempotents

In this section A will be a semisimple algebra over $\mathbb{C}$ which is not necessarily unital. (Topological considerations will not enter into our discussions. Our results apply equally well to a semiprime algebra over $\mathbb{C}$). In such an algebra the *socle* of A, soc(A) is defined to be the sum of the minimal right ideals (which equals the sum of the minimal left ideals (14) 30.10) or (O) if there are none. Definitions and theorems are usually stated for right ideals, corresponding statements may be made for left ideals.

F.1.1 DEFINITION. A non-zero idempotent $e \in A$ is *minimal* if eAe is a division algebra (if A is a Banach algebra then, by the Gelfand-Mazur theorem (14) 14.2, $eAe = \mathbb{C}e$). Min(A) denotes the set of minimal idempotents of A.

$\mathrm{Min}(\mathcal{B}(X))$ is the set of rank-one projections in $\mathcal{B}(X)$. If Ω is a compact Hausdorff space and $C(\Omega)$ denotes the algebra of continuous functions on Ω, the idempotents in $C(\Omega)$ are the characteristic functions of open and closed sets of Ω, while the minimal idempotents are the characteristic functions of isolated points of Ω. Thus $\mathrm{Min}(C(\Omega)) = \phi \iff \Omega$ possesses no isolated points.

F.1.2 DEFINITION. A right ideal R of A is *minimal* if $R \neq (O)$ and if for any right ideal $R_1 \subset R$, either $R_1 = (O)$, or $R_1 = R$.

The link between minimal idempotents and minimal ideals in semisimple algebras is set out in §BA.3, for reference it is restated here.

F.1.3 LEMMA. *Let* A *be semisimple then*

(i) $xA = O \iff x = O$;

(ii) R *is a minimal right ideal of* $A \iff R = eA$ *where* $e \in \mathrm{Min}(A)$;

(iii) *if* R *is a minimal right ideal of* A *and* $x \in A$ *then* xR *is either a minimal right ideal or is zero.*

Proof. (i) By (14) 24.17, $x \in \mathrm{rad}(A) = (O)$.
(ii), (iii) BA.3.1 and BA.3.2 ●

F.1.4 LEMMA (Exchange). *If* $\{e_1, \ldots e_n, f\} \subset \mathrm{Min}(A)$ *and* $f = e_1x_1 + \ldots + e_nx_n$ *where* $x_k \in A$ $(1 \leq k \leq n)$ *and* $e_jx_j \neq O$ *then*

$$\sum_k e_k A = fA + \sum_{k\neq j} e_k A.$$

Proof. It suffices to show that $e_j A$ is contained in the right hand side. Now $e_j x_j A = e_j A$ since $e_j x_j \neq 0$ and $e_j A$ is a minimal right ideal. Thus

$$e_j A = e_j x_j A \subset fA + \sum_{k\neq j} e_k A \quad \bullet$$

F.1.5 DEFINITION. A set W of idempotents of A is *orthogonal* if $ef = 0 = fe$ for $e, f \in W$ and $e \neq f$.

Observe that if $\{e_1, \ldots, e_n\}$ is an orthogonal set of idempotents in A, then

$$(e_1 + \ldots + e_n)A = e_1 A + \ldots + e_n A = e_1 A \oplus e_2 A \ldots \oplus e_n A.$$

We need two further technical lemmas.

F.1.6 LEMMA (Orthogonalisation). *Suppose that* $\{e_1, \ldots, e_n\}$ *is an orthogonal subset of* $\mathrm{Min}(A)$ *and that* $R = e_1 A + \ldots + e_n A$. *Then if* $f \in \mathrm{Min}(A)$ *is such that* $fA \not\subset R$, *there exists* e_{n+1} *such that* $\{e_1, \ldots, e_{n+1}\}$ *is an orthogonal subset of* $\mathrm{Min}(A)$ *and* $R + fA = e_1 A + \ldots + e_{n+1} A$.

Proof. Write $p = e_1 + \ldots + e_n$. Then $p^2 = p$ and $R = pA$. Since $fA \not\subset pA$ and $f = pf + (1 - p)f$, it follows that $(1 - p)f \neq 0$, hence $(1 - p)fA$ is a minimal right ideal which (F.1.3) contains a minimal idempotent g, say. Clearly $pg = 0$, hence if $e_{n+1} = g(1 - p)$, $e_{n+1} = e_{n+1}^2$. Further $e_{n+1} \neq 0$, for if $g = gp$, then $g = gpgp = 0$ which is false. Now $(0) \neq e_{n+1}A \subset gA$, hence $e_{n+1} \in \mathrm{Min}(A)$. Further $e_k e_{n+1} = e_k p e_{n+1} = 0 = e_{n+1} p e_k = e_{n+1} e_k$ $(1 \leq k \leq n)$, so the set $\{e_1, \ldots, e_{n+1}\}$ is an orthogonal subset of $\mathrm{Min}(A)$.

Now, since $(1 - p)fA = gA$ (F.1.3),

$$e_{n+1} = g(1 - p) \in gA = (1 - p)fA \subset fA + pA,$$

so $e_1A + \dots + e_{n+1}A \subset R + fA$.

Conversely, as $e_{n+1} = g(1 - p)$, by minimality (F.1.3)

$$(1 - p)fA = gA = g(1 - p)A = e_{n+1}A.$$

Therefore $fA \subset pA + e_{n+1}A$ giving $R + fA \subset \sum_1^{n+1} e_kA$ ●

F.1.7 LEMMA. *Let* R *be a right ideal of* A *lying in a finite sum of minimal right ideals of* A. *Then every orthogonal subset of* $R \cap \mathrm{Min}(A)$ *is finite, further, if* $\{e_1,\dots, e_n\}$ *is a maximal such subset then* $p = \sum_1^n e_k = p^2 \in \mathrm{soc}(A)$ *and* $R = pA$.

Proof. Using the exchange and orthogonalisation lemmas we can find an orthogonal set $\{f_1,\dots f_m\} \subset \mathrm{Min}(A)$ such that $R \subset f_1A + \dots + f_mA$. Suppose that $\{e_1,\dots, e_n\}$ is an orthogonal subset of $R \cap \mathrm{Min}(A)$. Then $e_1 = f_1a_1 + \dots + f_ma_m$ for some $a_k \in A$ $(1 \le k \le m)$, so $f_ja_j \ne 0$ for some j which we take to be 1, and then by the exchange lemma

$$\sum_1^m f_kA = e_1A + \sum_2^m f_kA.$$

Thus $e_2 = e_1b_1 + f_2b_2 + \dots + f_mb_m$ for some $b_k \in A$ $(1 \le k \le m)$, and since $e_2 \ne e_1b_1$ by orthogonality, $f_jb_j \ne 0$ for some j $(2 \le j \le m)$ which we take to be 2. Repeating the process m times we arrive at a stage at which $\sum_1^m f_kA = \sum_1^m e_kA$; at this point the process terminates since $R \subset \sum_1^m e_kA$, and if $e_{m+1} \in R \cap \mathrm{Min}(A)$ then $e_{m+1}e_k \ne 0$ for some k $(1 \le k \le m)$ so $\{e_1,\dots, e_{m+1}\}$ is not an orthogonal set. Hence no orthogonal subset of $R \cap \mathrm{Min}(A)$ contains more than m elements.

Suppose now that $\{e_1,\dots, e_n\}$ is a maximal orthogonal subset of $R \cap \mathrm{Min}(A)$ and put $p = \sum_1^n e_k$. Clearly $p^2 = p \in \mathrm{soc}(A)$ and $pA \subset R$, so it only remains to show that $R \subset pA$. If not, choose $w \in R \backslash pA$ and write $y = (1 - p)w$. Then $y \ne 0$, and since $y \in R \subset \sum_1^m f_kA$, $y = \sum_1^m f_ky$ so $f_ky \ne 0$ for some k $(1 \le k \le m)$. By minimality $f_kyA = f_kA$, hence $f_kyu = f_k$ for some $u \in A$. Then $yuf_k \ne 0$ so yuf_kA is a minimal right ideal which must contain an $f \in \mathrm{Min}(A)$. Now $fA \subset yA \subset (1 - p)A$, hence $fA \not\subset pA$ and, by F.1.6, there exists $e_{n+1} \in \mathrm{Min}(A)$ such that $\{e_1,\dots,e_{n+1}\}$

is orthogonal and $pA + fA = \sum_{1}^{n+1} e_k A$. However this implies that $e_{n+1} \in pA + fA \subset R$, contradicting the maximality of $\{e_1, \ldots, e_n\}$ which proves the lemma ●

F.1.8 DEFINITION. If $x \in A$ the *right annihilator* of x in A is defined by

$$\operatorname{ran}(x) = \{a \in A : xa = 0\};$$

while the *left annihilator* of x in A is defined by

$$\operatorname{lan}(x) = \{a \in A : ax = 0\}.$$

F.1.9 DEFINITION. If $x \in A$ we say that $p = p^2 \in A$ is a *left Barnes idempotent* for x in A if $xA = (1 - p)A$; while $q = q^2 \in A$ is a *right Barnes idempotent* for x in A if $Ax = A(1 - q)$.

Note that

(i) the Barnes idempotents, if they exist, are not normally unique;

(ii) p is a left Barnes idempotent for x in $A \Rightarrow \operatorname{lan}(x) = Ap$;

(iii) q is a right Barnes idempotent for x in $A \Rightarrow \operatorname{ran}(x) = qA$.

The next result is fundamental as it connects the existence of Barnes idempotents in the socle with left or right invertibility modulo the socle.

F.1.10 THEOREM (Barnes idempotents). *Let* A *be a unital semisimple algebra and* $x \in A$. *Then* x *is left (right) invertible modulo* $\operatorname{soc}(A) \Leftrightarrow x$ *has a right (left) Barnes idempotent in* $\operatorname{soc}(A)$.

Proof. Let u be a left inverse of x modulo $\operatorname{soc}(A)$ (the proof for right invertibility is similar). Then there exist $e_i \in \operatorname{Min}(A)$, $a_i \in A$ $(1 \le i \le n)$ such that

$$ux - 1 = \sum_{1}^{n} e_i a_i,$$

$$\Rightarrow \operatorname{ran}(x) \subset \sum_{1}^{n} e_i A,$$

$$\Rightarrow \operatorname{ran}(q) = qA, \text{ where } q = q^2 \in \operatorname{soc}(A) \quad (F.1.7).$$

Now $\quad ax = axq + ax(1-q) = ax(1-q)$,

hence $Ax \subset A(1-q)$. Observe that $Ax \cap Aq = (0)$. We shall show that $A = Ax \oplus Aq$ which implies that $Ax = A(1-q)$. Suppose not, then $Ax \oplus Aq$ is a proper left ideal of A which is contained in some maximal left ideal L. Since $ux - 1 \in soc(A)$ and $x \in L$ it follows that $soc(A) \not\subset L$ for, if it were, $1 \in L$ contradicting maximality. Thus there exists $g \in Min(A) \setminus L$. Now $L + Ag = A$ by maximality of L, and $L \cap Ag = (0)$ by minimality of Ag. So

$$A = L \oplus Ag.$$

In particular, for some $z \in L$, $a \in A$,

$$1 = z + ag,$$

hence, for $y \in L$,

$$y = yz + yag,$$

so $yag = y - yz \in L \cap Ag = (0)$. Hence

$$xag = 0 = qag.$$

Thus $ag \in ran(x) = qA$, so $ag = qag = 0$ which gives $1 = z \in L$, a contradiction which concludes the proof. The reverse inclusion is obviously true ●

F.1.11 EXAMPLE. We construct Barnes idempotents for a Fredholm operator T on a Banach space X. Recall that $F(X) = soc(\mathcal{B}(X))$. T is examined pictorially as in the proof of the Atkinson characterisation (0.2.2).

Let $X = ker(T) \oplus Z = W \oplus T(X)$ where Z and W are closed subspaces of X and $dim(ker(T)), dim(W) < \infty$. If $A \in \mathcal{B}(X)$,

$$TA = \begin{array}{c|c|c|} & \ker(T) & Z \\ \hline W & & \\ \hline T(X) & & T_{22} \\ \hline \end{array} \quad \begin{array}{c|c|c|} & \ker(T) & Z \\ \hline \ker(T) & A_{11} & A_{12} \\ \hline Z & A_{21} & A_{22} \\ \hline \end{array} = \begin{array}{c|c|c|} & \ker(T) & Z \\ \hline W & & \\ \hline T(X) & T_{22}A_{21} & T_{22}A_{22} \\ \hline \end{array} .$$

Since T_{22} is invertible, $A \in \operatorname{ran}(T) \iff A_{21} = 0 = A_{22}$. So if $A \in \operatorname{ran}(T)$

$$A = \begin{array}{c|c|c|} & \ker(T) & Z \\ \hline W & A_{11} & A_{12} \\ \hline T(X) & & \\ \hline \end{array} ,$$

and a right Barnes idempotent for T is

$$Q = \begin{array}{c|c|c|} & \ker(T) & Z \\ \hline \ker(T) & I & \\ \hline Z & & \\ \hline \end{array} ,$$

since $\operatorname{ran}(T) = Q\mathcal{B}(X)$. It is easily checked that $\mathcal{B}(X)T = \mathcal{B}(X)(I - Q)$. For if $B \in \mathcal{B}(X)$,

$$BT = \begin{array}{c|c|c|} & W & T(X) \\ \hline \ker(T) & B_{11} & B_{12} \\ \hline Z & B_{21} & B_{22} \\ \hline \end{array} \quad \begin{array}{c|c|c|} & \ker(T) & Z \\ \hline W & & \\ \hline T(X) & & T_{22} \\ \hline \end{array} = \begin{array}{c|c|c|} & \ker(T) & Z \\ \hline \ker(T) & & B_{12}T_{22} \\ \hline Z & & B_{22}T_{22} \\ \hline \end{array} ,$$

while if $C \in \mathcal{B}(X)$,

$$C(I - Q) = \begin{array}{c|c|c|} & \ker(T) & Z \\ \hline \ker(T) & C_{11} & C_{12} \\ \hline Z & C_{21} & C_{22} \\ \hline \end{array} \quad \begin{array}{c|c|c|} & \ker(T) & Z \\ \hline \ker(T) & & \\ \hline Z & & I \\ \hline \end{array} = \begin{array}{c|c|c|} & \ker(T) & Z \\ \hline \ker(T) & & C_{12} \\ \hline Z & & C_{22} \\ \hline \end{array} .$$

Since T_{22} is invertible the equations $B_{12}T_{22} = C_{12}$, $B_{22}T_{22} = C_{22}$ may be solved uniquely for B_{12}, B_{22} which verifies the equality. Observe that Z is an arbitrary closed complement of $\ker(T)$ in X so a right Barnes idempotent for T is any idempotent in $\mathcal{B}(X)$ with range $\ker(T)$.

A similar analysis shows that any idempotent P in $\mathcal{B}(X)$ whose kernel is $T(X)$ satisfies

$$\text{lan}(T) = \mathcal{B}(X)P, \quad T\mathcal{B}(X) = (1 - P)\mathcal{B}(X),$$

and hence is a left Barnes idempotent for T.

F.2 Primitive Banach algebras

In this section A will be a primitive unital Banach algebra over $\mathbb{C}$ such that $\text{Min}(A) \neq \emptyset$. Recall that an algebra A is *primitive* if $(0) \in \Pi(A)$, that is, A possesses a faithful irreducible representation. Note that, if A is a primitive algebra, then

$$x, y \in A \text{ and } xAy = (0) \Rightarrow \text{either } x \text{ or } y \text{ is zero.}$$

For suppose that $x \neq 0 \neq y$ and let ψ be a faithful irreducible representation of A on the linear space X. Then there exist $\zeta, \eta \in X$ such that $\psi(x)\zeta \neq 0 \neq \psi(y)\eta$. Now $\psi(Ay)\eta$ is a subspace of X which is invariant under each element of $\psi(A)$, hence $\psi(Ay)\eta = X$. So there exists $z \in A$ such that $\psi(zy)\eta = \zeta$. But then $\psi(xzy)\eta = \psi(x)\zeta \neq 0$ so $xAy \neq (0)$.

If X and Y are two linear spaces we write $\dim(X) = \dim(Y)$ to mean that either the spaces are both infinite dimensional or they have the same finite dimension.

F.2.1 LEMMA. *Let* $e, f \in \text{Min}(A)$ *and* R *be a right ideal of* A, *then*

(i) *there exist* $u, v \in A$ *such that* $f = uev$;

(ii) $\dim(eAf) = 1$;

(iii) $\dim(Re) = \dim(Rf)$;

(iv) $\dim(Ae/Re) = \dim(Af/Rf)$.

<u>Proof</u>. (i) We have observed that $eAf \neq (0)$. Choose a non-zero $v \in eAf$. Since Af is a minimal left ideal, $Af = Av$, so $f = uv$ for some $u \in A$.

Also $v = ev$, hence $f = uev$.

(ii) By (i), $eAf = eAuev \subset eAev = \mathbb{C}ev$ whose dimension is unity.

(iii) By (i), $Rf = Ruev \subset Rev$. So if $\dim(Re) < \infty$, so is $\dim(Rev)$ and $\dim(Rf) \leq \dim(Re)$. Similarly, if $\dim(Rf) < \infty$, then $\dim(Re) \leq \dim(Rf)$. Further if Rf is infinite dimensional so is Rev and therefore so is Re, and conversely.

(iv) Let S be a subset of A, then

Sf is linearly independent modulo Rf,

$\Rightarrow$ $Suev$ is linearly independent modulo $Ruev$, by (i),

$\Rightarrow$ Sue is linearly independent modulo $Rue = Re$.

(To see this since $ue \neq 0$, $Aue = Ae$ so $tue = e$ for some $t \in A$. Hence $Re = Rtue \subset Rue \subset Re$). It follows that $\dim(Ae/Re) < \infty \Rightarrow \dim(Af/Rf) \leq \dim(Ae/Re)$; similarly $\dim(Af/Rf) < \infty \Rightarrow \dim(Ae/Re) \leq \dim(Af/Rf)$. The infinite dimensional results are clear ●

For the remainder of this section e will denote a fixed minimal idempotent of A. We shall write

$$x \to \hat{x} : A \to \mathcal{B}(Ae)$$

to denote the left regular representation of the primitive Banach algebra A on the Banach space Ae, that is $\hat{x}(y) = xy$ for $y \in Ae$. This is an important representation. It is norm reducing, hence continuous and since $xAe = 0 \Rightarrow x = 0$, it is faithful. Further if L is a subspace of Ae which is invariant under $\hat{x}$, for each $x \in A$, it follows that L is a left ideal of A which implies that either $L = 0$ or $L = Ae$. Thus the representation is irreducible.

Observe that

$$\hat{x}(Ae) = xAe,$$

and $\ker(\hat{x}) = \mathrm{ran}(x) \cap Ae = \mathrm{ran}(x)e$.

Since xA and $\mathrm{ran}(x)$ are right ideals of A, it follows from F.2.1 that the rank, nullity and defect of the operator $\hat{x} \in \mathcal{B}(Ae)$ are independent of the particular choice of $e \in Min(A)$. As the following example illustrates we can say even more when dealing with the algebra of bounded linear operators on a Banach space.

F.2.2 EXAMPLE. Let X be a Banach space and let A be any unital closed subalgebra of $\mathcal{B}(X)$ containing $\mathcal{F}(X)$. Then A is a primitive Banach algebra and we fix our minimal idempotent to be the rank-one projection $y \otimes g$ where $y \in X$, $g \in X^*$ and $g(y) = 1$. The representation space is now

$$A(y \otimes g) = Ay \otimes g = X \otimes g,$$

since A contains all rank-one operators. Then, if $T \in A$,

$$\hat{T}(x \otimes g) = Tx \otimes g$$

defines the corresponding $\hat{T} \in \mathcal{B}(X \otimes g)$. It is clear that the rank, nullity and defect of $\hat{T}$ are equal to those of T as an operator on X.

F.2.3 DEFINITION. If $x \in A$ we define the *rank* of x by $\text{rank}(x) = \text{rank}(\hat{x})$.

F.2.4 THEOREM. (i) $x = 0 \iff \text{rank}(x) = 0$;

(ii) $soc(A) = \{x \in A : \text{rank}(x) < \infty\}$.

Proof. (i) is obvious.

(ii) As an inductive hypothesis assume that $\dim(xAe) \leq n \Rightarrow x \in soc(A)$. Note that $n = 0 \Rightarrow x = 0$. Suppose, then, that $\dim(xAe) \leq n + 1$. Since A is primitive there exists $u \in A$ such that $xue \neq 0$. Then $xueA$ is a minimal right ideal which therefore contains an $f \in Min(A)$. Now there exists a non-zero $v \in fAe \subset xAe$. Since $(1 - f)v = 0$ it follows that $\dim((1 - f)xAe) \leq n$, hence, by the induction hypothesis, $x - fx \in soc(A)$. Therefore $x \in soc(A)$ which completes the proof.

Conversely, if $x \in soc(A)$, $xA \subset \sum_1^n f_i A$ where $f_i \in Min(A)$ $(1 \leq i \leq n)$ hence $xAe \subset \sum_1^n f_i Ae$ and $\dim(xAe) \leq n$ by F.2.1 ●

F.2.5 DEFINITION. x is defined to be a *Fredholm* element of A if there exists a $y \in A$ such that $xy - 1$, $yx - 1 \in soc(A)$. The set of Fredholm elements of A is written $\Phi(A)$.

If soc(A) is a proper ideal of A then $x \in \Phi(A) \Longleftrightarrow x$ is invertible modulo soc(A). (We often implicitly make this assumption, otherwise $A = \Phi(A)$). By BA.2.4 invertibility modulo soc(A) is equivalent to invertibility modulo k(h(soc(A))). Since the latter ideal is closed in A it follows that $\Phi(A)$ is an open multiplicative semigroup of A which is stable under perturbations by elements of k(h(soc(A))).

Next we link Fredholm elements in A with Fredholm operators on Ae.

F.2.6 THEOREM. $x \in \Phi(A) \Rightarrow \hat{x}$ *is a Fredholm operator on* Ae.

<u>Proof.</u> We use the Barnes idempotents p, q of F.1.9 and F.1.10.

$$\ker(\hat{x}) = \operatorname{ran}(x) \cap Ae = qA \cap Ae = qAe,$$

therefore $n(\hat{x}) = \operatorname{rank}(\hat{q}) = \operatorname{rank}(q) < \infty$ (F.2.4).

$$\hat{x}(Ae) = xAe = (1 - p)Ae \text{ which is closed in } Ae, \text{ and}$$

$$Ae/\hat{x}(Ae) = Ae/(1 - p)Ae \cong pAe,$$

so $d(\hat{x}) = \operatorname{rank}(\hat{p}) = \operatorname{rank}(p) < \infty$ (F.2.4) ●

Example F.4.2 shows that the converse of this theorem is false.

F.2.7 DEFINITION. If $x \in \Phi(A)$ we define the *nullity*, *defect* and *index* of x by

$$\operatorname{nul}(x) = n(\hat{x}), \quad \operatorname{def}(x) = d(\hat{x}), \quad \operatorname{ind}(x) = i(\hat{x}).$$

Now if $x \in \Phi(A)$, $\hat{x}$ is a Fredholm operator on Ae with the same nullity, defect and index; further, nul(x) = rank(q) and def(x) = rank(p) where q, p are right and left Barnes idempotents for x in A, respectively.

We are now in a position to reap the benefits of the connection between the Fredholm theories in a primitive Banach algebra with minimal ideals and on a Banach space.

F.2.8 THEOREM. $\mathrm{Inv}(A) = \{x \in \Phi(A) : \mathrm{nul}(x) = 0 = \mathrm{def}(x)\}$.

Proof. $x \in \mathrm{Inv}(A)$,

$\iff xA = A = Ax$,

$\iff x \in \Phi(A)$ and the Barnes idempotents $p = q = 0$,

$\iff x \in \Phi(A)$ and $\mathrm{nul}(x) = \mathrm{rank}(p) = 0 = \mathrm{rank}(q) = \mathrm{def}(x)$ ●

F.2.9 THEOREM (Index).

(i) *The map* $x \to \mathrm{ind}(x) : \Phi(A) \to Z$ *is continuous;*

(ii) $\mathrm{ind}(xy) = \mathrm{ind}(x) + \mathrm{ind}(y)$, $(x, y \in \Phi(A))$;

(iii) $\mathrm{ind}(x) = \mathrm{ind}(y)$ *if* x *and* y *lie in the same component of* $\Phi(A)$;

(iv) $\mathrm{ind}(x) = \mathrm{ind}(x + u)$, $(x \in \Phi(A),\ u \in k(h(\mathrm{soc}(A))))$.

Proof. (i) The map $x \to \hat{x}$ is continuous. Now use the continuity of the index for Fredholm operators.

(ii) $\mathrm{ind}(xy) = i(\widehat{xy}) = i(\hat{x})i(\hat{y}) = i(\hat{x}) + i(\hat{y}) = \mathrm{ind}(x) + \mathrm{ind}(y)$.

(iii) if x and y are connected by a path in $\Phi(A)$ the same is true of $\hat{x}$ and $\hat{y}$ in the Fredholm operators on Ae.

(iv) x and u are connected by the path $\{x + \lambda u : 0 \leq \lambda \leq 1\}$ which lies in $\Phi(A)$. Now use (iii) ●

F.2.10 THEOREM (Punctured neighbourhood). *If* $x \in \Phi(A)$ *there exists* $\varepsilon > 0$ *such that*

(i) $\mathrm{nul}(x + \lambda)$ *is a constant* $\leq \mathrm{nul}(x)$, $(0 < |\lambda| < \varepsilon)$;

(ii) $\mathrm{def}(x + \lambda)$ *is a constant* $\leq \mathrm{def}(x)$, $(0 < |\lambda| < \varepsilon)$;

(iii) $\mathrm{ind}(x + \lambda)$ *is a constant*, $(|\lambda| < \varepsilon)$.

Proof. Apply the punctured neighbourhood theorem for Fredholm operators O.2.7 ●

F.2.11 THEOREM. *If* $x \in \Phi(A)$ *and* $\mathrm{ind}(x) \leq 0$ (≥ 0) *there exists* $u \in \mathrm{soc}(A)$ *such that* $x + \lambda u$ *is left (right) invertible for* $\lambda \neq 0$. *We may choose* $u \in pAq$ *where* p *and* q *are left and right Barnes idempotents for* x *in* A *(respectively).*

Proof. We consider first the case of $\mathrm{ind}(x) = 0$. Let p and q be left and right Barnes idempotents for x in A, then $Ax = A(1 - q)$, $xA = (1 - p)A$, hence there exist $y, z \in A$ such that $yx = 1 - q$, $xz = 1 - p$. Because $px = xq = 0$, we may take $y \in A(1 - p)$, $z \in (1 - q)A$.

Since $\mathrm{ind}(x) = 0$, the spaces qAe and pAe have the same finite dimension. So by the Jacobson density theorem ((75) 2.4.16) there exist $s, t \in A$ such that $\hat{s}(\in \mathcal{B}(Ae))$ takes a basis of qAe onto a basis of pAe while $\hat{t}(\in \mathcal{B}(Ae))$ reverses the process. Write $u = psq$, $v = qtp$ then

$$\hat{s}(qae) = sqae = psqae; \quad \hat{u}(qae) = psqae,$$

so $\hat{u}|qAe = \hat{s}|qAe$. Similarly $\hat{v}|pAe = \hat{t}|pAe$.

Now $\hat{u}\hat{v}(ae) = uvae = psqtpae = pstpae = pae,$

and $\hat{v}\hat{u}(ae) = vuae = qtpsqae = qtsqae = qae,$

by choice of $\hat{s}$ and $\hat{t}$. So $\hat{u}\hat{v} = \hat{p}$, $\hat{v}\hat{u} = \hat{q}$. Hence $uv = p$, $vu = q$ as the representation $x \to \hat{x}$ is faithful.

Now $yu = 0 = vx$, hence for $\lambda \neq 0$,

$$(y + \lambda^{-1}v)(x + \lambda u) = 1 - q + q = 1.$$

Thus $x + \lambda u$ has a left inverse in A. Similarly $uz = 0 = xv$ so

$$(x + \lambda u)(z + \lambda^{-1}v) = 1 - p + p = 1,$$

hence $x + \lambda u$ has a right inverse and is therefore in $\mathrm{Inv}(A)$ for $\lambda \neq 0$.

For the case $\mathrm{ind}(x) \leq 0$ let $\{c_1, \ldots, c_n\}$ be a basis for qAe and $\{d_1, \ldots, d_m\}$ a basis for pAe where $n \leq m$. Now use the Jacobson density theorem to construct $s, t \in A$ such that

$$\hat{t}(d_i) = c_i \quad (1 \le i \le n),$$

$$\hat{s}(c_i) = d_i \quad (1 \le i \le n).$$

Then $\hat{t}\hat{s}$ is the identity map restricted to qAe and the same argument gives $x + \lambda u$ left-invertible in A for $\lambda \neq 0$. A similar proof deals with the case of $\mathrm{ind}(x) \geq 0$ ●

Note that this method of developing Fredholm theory in a primitive Banach algebra A requires that $\mathrm{Min}(A) \neq \emptyset$. But $\mathrm{Min}(A) = \emptyset \iff \mathrm{soc}(A) = (0)$ and then Fredholm theory is trivial for $\Phi(A) = \mathrm{Inv}(A)$ and obviously the nullity, defect and index of any Fredholm element (however these concepts are defined) must be zero. The Calkin algebra of a separable Hilbert space is an example of a primitive Banach algebra with zero socle.

F.3 General Banach algebras

We now extend our theory to a unital Banach algebra A using its primitive quotient algebras as building blocks. The quotient algebra $A' = A/\mathrm{rad}(A)$ is semisimple and we write x' for the coset $x + \mathrm{rad}(A)$ and if $S \subset A$ write $S' = \{x' : x \in S\}$. In general the socle of A does not exist so in its place we use the presocle.

F.3.1 DEFINITION. The *presocle* of A is defined by

$$\mathrm{psoc}(A) = \{x \in A : x' \in \mathrm{soc}(A')\}.$$

Clearly $\mathrm{psoc}(A)$ is an ideal of A, while if A is semisimple, $\mathrm{psoc}(A) = \mathrm{soc}(A)$. The ideal of *inessential elements* of A is defined to be

$$I(A) = k(h(\mathrm{psoc}(A))).$$

Clearly $I(A)$ is a closed ideal of A.

F.3.2 DEFINITION. An element x is called a *Fredholm element* of A if there exists $y \in A$ such that $xy - 1,\ yx - 1 \in \mathrm{psoc}(A)$. The set of

Fredholm elements of A is written $\Phi(A)$.

If $psoc(A)$ is a proper ideal of A then $x \in \Phi(A) \iff x$ is invertible modulo $psoc(A)$, while if $psoc(A) = A$ then $A = \Phi(A)$. At the other extreme $psoc(A) = rad(A) \iff soc(A') = (O) \iff \Phi(A) = Inv(A)$ (BA.2.2).

By BA.2.4 invertibility modulo $soc(A)$ is equivalent to invertibility modulo $I(A)$. Thus $\Phi(A)$ is an open semigroup of A which is stable under perturbations by elements of $I(A)$. Note that by BA.2.2 and BA.2.5 we have

$$Inv(A)' = Inv(A'); \quad \Phi(A)' = \Phi(A'); \quad I(A)' = I(A').$$

If $P \in \Pi(A)$ then A/P is a primitive unital Banach algebra and this fact enables us to develop Fredholm theory in A. Further, by BA.2.3, the structure space of $A/I(A)$ is isomorphic to $h(I(A)) = h(psoc(A))$. Thus using BA.2.2 we get

$$\begin{aligned} & x \in \Phi(A), \\ \iff & x \text{ is invertible modulo } I(A), \\ \iff & x \text{ is invertible modulo } P \quad (P \in h(soc(A))). \end{aligned}$$

F.3.3 LEMMA. *If* $s' \in Min(A')$ *there exists a unique* $P \in \Pi(A)$ *such that* $s \notin P$, *further* $s + P \in Min(A/P)$.

<u>Proof.</u> The first statement follows from BA.2.5 and BA.3.5. Now since $s \notin P$, $s + P \neq 0$, so if $a \in A$

$$(s + P)(a + P)(s + P) = sas + P.$$

But $s' \in Min(A')$ so $s'a's' = \lambda s'$ for some $\lambda \in \mathbb{C}$, hence $sas - \lambda s \in rad(A) \subset P$, so

$$(s + P)(a + P)(s + P) = \lambda(s + P),$$

that is $s + P \in \mathrm{Min}(A/P)$ ●

F.3.4 THEOREM. *If* $x \in \Phi(A)$ *there exist* $\varepsilon > 0$ *and a finite subset* Ω *of* $\Pi(A)$ *such that if* $y \in A$ *and* $||x - y|| < \varepsilon$ *then*

(i) $y + P \in \Phi(A/P)$, $(P \in \Omega)$;

(ii) $y + P \in \mathrm{Inv}(A/P)$, $(P \in \Pi(A)\setminus\Omega)$.

Proof. If $\mathrm{Min}(A') = \emptyset$ then $\Phi(A) = \mathrm{Inv}(A)$ and the theorem is trivial so suppose that this is not the case. Then if $x \in \Phi(A)$ there exist elements u; a_i, s_i $(1 \leq i \leq \ell)$; b_j, t_j $(1 \leq j \leq m)$ in A such that s'_i, $t'_j \in \mathrm{Min}(A')$ for each i, j and

$$ux - 1 - \sum_1^{\ell} s_i a_i \in \mathrm{rad}(A),$$

$$xu - 1 - \sum_1^{m} t_j b_j \in \mathrm{rad}(A).$$

By F.3.3 the set $h(\{s_1, \ldots s_\ell; t_1, \ldots t_m\})$ has a finite complement Ω in $\Pi(A)$. Now

$$J = k(h(\{s_1, \ldots s_\ell; t_1, \ldots t_m\}))$$

is a closed ideal of A and

$$ux + J = 1 + J = xu + J,$$

that is x is invertible modulo J. Hence there exists $\varepsilon_o > 0$ such that $||x - y|| < \varepsilon_o \Rightarrow y$ is invertible modulo J. Now $J \subset P$ for $P \in \Pi(A)\setminus\Omega$, so $||x - y|| < \varepsilon_o \Rightarrow y$ is invertible modulo P for $P \in \Pi(A)\setminus\Omega$ proving (ii).

If $P \in \Pi(A)$ then

$$u'x' - 1', x'u' - 1' \in \mathrm{soc}(A'),$$

$$\Rightarrow \quad u'x' - 1' + P', x'u' - 1' + P' \in \mathrm{soc}(A'/P'), \quad \text{(BA.3.4)}$$

$\Rightarrow \quad ux - 1 + P,\ xu - 1 + P \in soc(A/P)$, (BA.2.6)

$\Rightarrow \quad x + P \in \Phi(A/P)$.

Now if $\Omega = \{P_1, \ldots, P_n\}$, then there exists $\varepsilon_k > 0$ such that $||x - y|| < \varepsilon_k \Rightarrow y + P_k \in \Phi(A/P_k)$ $(1 \leq k \leq n)$. A choice of $\varepsilon = \min\{\varepsilon_o, \varepsilon_1, \ldots \varepsilon_k\}$ gives (i) ●

Recall that if, in a primitive algebra A, Min(A) = ∅ then the nullity and defect of every Fredholm element are defined to be zero.

F.3.5 DEFINITION. For each $x \in \Phi(A)$ we define the *nullity, defect* and *index functions* $\Pi(A) \to \mathbb{Z}$ by

$$\nu(x)(P) = \mathrm{nul}(x + P),$$

$$\delta(x)(P) = \mathrm{def}(x + P),$$

$$\iota(x)(P) = \mathrm{ind}(x + P).$$

By F.3.4 each of these functions has finite support in $\Pi(A)$ and is zero on h(psoc(A)). Obviously $\iota(x) \equiv \nu(x) - \delta(x)$.

If A is a primitive Banach algebra then, for each minimal idempotent, (O) is the unique minimal ideal which fails to contain it (F.3.5). So if $(O) \neq P \in \Pi(A)$, $\mathrm{Min}(A) \subset P$ hence $\mathrm{soc}(A) \subset P$. It follows that, in this case the support of the nullity, defect and index functions consists of the zero ideal, so

$$\mathrm{nul}(x) = \nu(x)(O), \quad \mathrm{def}(x) = \delta(x)(O), \quad \mathrm{ind}(x) = \iota(x)(O).$$

The concepts of *nullity, defect* and *index* can be extended to a general Banach algebra as follows.

F.3.6 DEFINITION. If $x \in \Phi(A)$ we define

$$\mathrm{nul}(x) = \sum_{P \in \Pi(A)} \nu(x)(P),$$

$$\mathrm{def}(x) = \sum_{P \in \Pi(A)} \delta(x)(P),$$

$$\mathrm{ind}(x) = \sum_{P \in \Pi(A)} \iota(x)(P).$$

Since $\nu(x)(P) \geq 0$, $\delta(x)(P) \geq 0$ it follows that

$$\mathrm{nul}(x) = 0 \iff \nu(x) \equiv 0,$$

$$\mathrm{def}(x) = 0 \iff \delta(x) \equiv 0.$$

Now F.2.8 extends to the general case.

F.3.7 THEOREM.
$\mathrm{Inv}(A) = \{x \in \Phi(A) : \mathrm{nul}(x) = 0 = \mathrm{def}(x)\} = \{x \in \Phi(A) : \nu(x) \equiv 0 \equiv \delta(x)\}$.

Proof. Apply BA.2.2(v) ●

The properties of the index of a Fredholm element in a primitive Banach algebra given in F.2.9 extend easily to the index function.

F.3.8 THEOREM (Index).

(i) *The map* $x \to \iota(x) : \Phi(A) \to \mathbb{Z}^{\Pi(A)}$ *is continuous in the pointwise topology on* $\mathbb{Z}^{\Pi(A)}$.

(ii) $\iota(xy) = \iota(x) + \iota(y)$, $(x, y \in \Phi(A))$;

(iii) $\iota(x) = \iota(y)$ *if* x *and* y *lie in the same component of* $\Phi(A)$.

(iv) $\iota(x + u) = \iota(x)$, $(x \in \Phi(A),\ u \in I(A))$.

F.3.9 THEOREM (Punctured neighbourhood). *Fix* $x \in \Phi(A)$, *then there exists* $\varepsilon > 0$ *such that for each* $P \in \Pi(A)$,

(i) $\nu(x + \lambda)(P)$ *is a constant* $\leq \nu(x)(P)$, $(0 < |\lambda| < \varepsilon)$;

(ii) $\delta(x + \lambda)(P)$ *is a constant* $\leq \delta(x)(P)$, $(0 < |\lambda| < \varepsilon)$;

(iii) $\iota(x + \lambda)(P)$ *is a constant*, $(|\lambda| < \varepsilon)$.

Proof. Choose ε as in F.3.3 and label it ε_o. Then by F.3.4 and the punctured neighbourhood theorem in primitive algebras (F.2.10) there exist positive numbers $\varepsilon_1, \ldots, \varepsilon_n$ such that

$$\nu(x + \lambda)(P_k) \quad = \quad \text{nul}(x + \lambda + P_k) \quad \text{which is a constant} \; \leq$$

$$\leq \; \nu(x)(P_k) \quad = \quad \text{nul}(x + P_k) \quad \text{for} \quad 0 < |\lambda| < \varepsilon_k \quad (1 \leq k \leq n).$$

Take $\varepsilon = \min\{\varepsilon_o, \varepsilon_1, \ldots, \varepsilon_n\}$ and the result for $\nu(x)$ follows. The proof for $\delta(x)$ is similar ●

We remark that in a general Banach algebra if $x \in \Phi(A)$ then $\iota(x) \equiv 0 \Rightarrow \text{ind}(x) = 0$ but the converse is not necessarily true. This fact has important consequences which were first observed by Pearlman (69). If T is an index-zero Fredholm operator on a Banach space then we have the following important decomposition (0.2.8), $T = V + F$ where V is invertible and F of finite rank (with analogous results if $i(T) \leq 0$ or ≥ 0). The converse is obviously true and by F.3.7 and F.3.8 it extends to general Banach algebras. However, as the next example shows, if $x \in \Phi(A)$ and $\text{ind}(x) = 0$ it does not follow that x has a corresponding decomposition.

F.3.10 EXAMPLE. Let H_1, H_2 be infinite dimensional separable Hilbert spaces and take $A = \mathcal{B}(H_1) \oplus \mathcal{B}(H_2)$. Then A is a semisimple Banach algebra and $\text{soc}(A) = \mathcal{F}(H_1) \oplus \mathcal{F}(H_2)$. Considering Fredholm theory in A relative to the socle, if $T \in \Phi(A)$, $T = T_1 \oplus T_2$ and $\text{ind}(T) = i_{H_1}(T_1) + i_{H_2}(T_2)$.
Let $S = U_1 \oplus V_2$ where U_1 is the forward unilateral shift on H_1 and V_2 the backward unilateral shift on H_2. Clearly $\text{ind}(S) = 0$. Suppose that there exists $F \in \text{soc}(A)$ such that $S + F \in \text{Inv}(A)$. Then $F = F_1 \oplus F_2$, where $F_1 \in \mathcal{F}(H_1)$, $F_2 \in \mathcal{F}(H_2)$ and $S + F = (U_1 + F_1) \oplus (V_2 + F_2) \in \text{Inv}(A)$ $\Leftrightarrow U_1 + F_1 \in \text{Inv}\,\mathcal{B}(H_1)$ and $V_2 + F_2 \in \text{Inv}(\mathcal{B}(H_2))$.
But $i_{H_1}(U_1 + F_1) = i_{H_1}(U_1) = -1$ while $i_{H_2}(V_2 + F_2) = i_{H_2}(V_2) = +1$ which is impossible.

To overcome this difficulty we employ the index function. The representations $S_1 \oplus S_2 \to S_1$ and $S_1 \oplus S_2 \to S_2$ are clearly irreducible therefore $P_1 = \mathcal{B}(H_1) \oplus (0)$, $P_2 = (0) \oplus \mathcal{B}(H_2)$ are primitive ideals of A which do not contain $\text{soc}(A)$. Suppose $P \subset \Pi(A)$, $\text{soc}(A) \not\subset P$, then there exists

$E \in Min(A)$ such that $E \not\subseteq P$. Since $0 \neq E = E_1 \oplus E_2$, either E_1 or E_2 are non zero. If $E_1 \neq 0$, $E \not\subseteq P_2$ so by BA.3.5, $P = P_2$, similarly if $E_2 \neq 0$, $P = P_1$. Thus we have shown that P_1 and P_2 are the only two primitive ideals which do not contain soc(A). So for $T \in \Phi(A)$ and $P \neq P_1$ or P_2, $\iota(T)(P) = 0$. On the other hand if $T = T_1 \oplus T_2$ $\iota(T)(P_1) = i_{H_1}(T_1)$; $\iota(T)(P_2) = i_{H_2}(T_2)$. It is now easy to see that if $\iota(T) \equiv 0$ then $i_{H_1}(T_1) = 0$ and $i_{H_2}(T_2) = 0$ so, by applying O.2.8 to T_1 and T_2, $T = V + F$ where $V \in Inv(A)$ and $F \in soc(A)$.

This idea can be made precise.

F.3.11 THEOREM. *If* $x \in \Phi(A)$ *and* $\iota(x)(P) \leq 0$ (≥ 0) *for all* $P \in \Pi(A)$ *there exists* $u \in I(A)$ *such that* $x + \lambda u$ *is left(right) invertible for* $\lambda \neq 0$.

Proof. We consider the case $\iota(x) \equiv 0$, the remaining cases may be handled as in F.2.11.

Let $x \in \Phi(A)$ then x' is invertible modulo $soc(A')$. Thus there exist $p, q \in A$ such that p' and q' are left and right Barnes idempotents for x' in A' (F.1.10), in particular $x'A' = (1' - p')A'$. Since $p' \in soc(A')$ there exist $s_1, \ldots, s_n \in A$ such that $\{s'_1, \ldots, s'_n\}$ is an orthogonal subset of $Min(A')$ and

$$p = (s_1 + \ldots + s_n)p \text{ modulo } rad(A),$$

hence $p = (s_1 + \ldots + s_n)p$ modulo P $(P \in \Pi(A))$.

Thus $xA = (1 - (s_1 + \ldots + s_n)p)A$ modulo P,

so if $\{s_1, \ldots, s_n\} \subset P$, which is true for all but a finite subset $\{P_1, \ldots, P_m\}$ of $\Pi(A)$, it follows that x is right invertible modulo P except for $P = P_k$ $(1 \leq k \leq m)$. Since $\iota(x)(P) = 0$ $(P \in \Pi(A))$ we deduce that x is invertible modulo P for $P \neq P_k$ $(1 \leq k \leq m)$.

Now for $1 \leq k \leq m$, by F.2.11 we may choose $t_k \in A$ such that

$$t_k \in pA = (s_1 + \ldots + s_n)pA \text{ modulo } P_k,$$

with $x + \lambda t_k$ invertible modulo P_k for $\lambda \neq 0$.

Put $u_k = \Big(\sum_{s_i \notin P_k} s_i\Big) t_k$, $(1 \leq k \leq m)$.

Now $s'_i \in soc(A') \Rightarrow s_i \in I(A)$ for each i, hence $u_k \in I(A)$.

Then
$$x + \lambda u_k = x + \lambda\Big(\sum_{s_i \notin P_k} s_i\Big)\Big(\sum_1^n s_i\Big)pt_k \text{ modulo } P_k,$$
$$= x + \lambda\Big(\sum_{s_i \notin P_k} s_i\Big)pt_k \text{ modulo } P_k,$$
$$= x + \lambda t_k \text{ modulo } P_k$$

which is invertible modulo P_k for $\lambda \neq 0$. Further u_k lies in every primitive ideal except P_k. So

$$x = x + \lambda u_k \text{ modulo } P \text{ for } P \neq P_k.$$

Write $u = \sum_1^m u_k$. Then

$$x + \lambda u = x + \lambda u_k \text{ modulo } P_k$$

which is invertible modulo P_k for $\lambda \neq 0$ and $1 \leq k \leq m$, while $x + \lambda u = x$ modulo P for $P \neq P_k$ $(1 \leq k \leq m)$. Thus, for $\lambda \neq 0$, $x + \lambda u$ is invertible modulo P for $P \in \Pi(A)$. It follows by BA.2.2 that $x + \lambda u \in Inv(A)$ for $\lambda \neq 0$ ●

A final generalisation of our theory remains.

F.3.12 DEFINITION. An ideal K of A such that $K \subset I(A)$ is called an *inessential ideal* of A. An $x \in A$ such that there exists $y \in A$ such that $xy - 1, yx - 1 \in K$ is called a *K-Fredholm* element of A. The set of K-Fredholm elements is denoted by $\Phi_K(A)$.

We can develop a Fredholm theory relative to each such K and, by (BA.2.4), without loss of generality we can assume K to be norm closed or to equal $k(h(K))$. The statements and proofs all go through with only the obvious modifications. An inessential ideal of particular importance is the algebraic kernel which is considered in §F.4.

F.4 Notes

Fredholm theory in an algebraic setting was pioneered by Barnes (7), (8), in 1968, 9 in the context of a *semiprime* ring (one possessing no non-zero nilpotent left or right ideals). He used the concept of an ideal of finite order to replace the finite dimensionality of the kernel and co-range of a Fredholm operator.

F.4.1 DEFINITION. A right ideal J in a semiprime ring A has *finite order* if it is contained in a finite sum of minimal right ideals of A (with a corresponding definition on the left). The *order* of an ideal J, written $\mathrm{ord}(J)$, is defined to be the smallest number of minimal ideals whose sum is J.

The connection with our work is clear, for if $x \in \Phi(A)$ and p and q are left and right Barnes idempotents, then the left ideal $\mathrm{lan}(x) = Ap$, and the right ideal $\mathrm{ran}(x) = qA$, both have finite order so the nullity, defect and index of x are defined by the formulae

$$\mathrm{nul}(x) = \mathrm{ord}(\mathrm{ran}(x)) = \mathrm{ord}(qA),$$

$$\mathrm{def}(x) = \mathrm{ord}(\mathrm{lan}(x)) = \mathrm{ord}(Ap),$$

$$\mathrm{ind}(x) = \mathrm{nul}(x) - \mathrm{def}(x).$$

If A is primitive and $x \in \Phi(A)$ then $\mathrm{ord}(qA) = \mathrm{rank}(q)$ and $\mathrm{ord}(Ap) = \mathrm{rank}(p)$ so the definition of these concepts coincides with our own.

The index theory which Barnes obtains is more general than that developed in Chapter F as it is purely algebraic in character, but each result must be proved ab initio, and the preliminary manipulations are rather involved. Our approach, developed by Smyth, via the left regular representation of a primitive algebra A on Ae where $e \in \mathrm{Min}(A)$, and the link between Fredholm elements in A and Fredholm operators on Ae (F.2.6) is more direct. However our theory is less general than that of Barnes, for F.2.1(ii) requires that A be a Banach algebra.

The representation which we have used is well known ((75) 2.4.16), the correspondence between the dimensions of the kernel and the co-range of x and $\hat{x}$ are the key to our exposition of Fredholm theory. We now give an example of a primitive Banach algebra A with $\mathrm{Min}(A) \neq \emptyset$ and an $x \notin \Phi(A)$ such that $\hat{x}$ is a Fredholm operator on Ae showing that the converse of F.2.6 is false in general.

F.4.2 EXAMPLE. Let T be an operator on a Banach space X such that

$$\sigma(T) = \omega(T) = \sigma_{\mathcal{B}(X)/\mathcal{K}(X)}(T + \mathcal{K}(X)) = \{\lambda : |\lambda| = 1\}.$$

(The bilateral shift on a separable Hilbert space is an example). Take A to be the closed unital subalgebra of $\mathcal{B}(X)$ generated by T and $\mathcal{K}(X)$. Then A is a primitive Banach algebra with $\mathrm{Min}(A) \neq \emptyset$ and, as in F.2.2, the rank, nullity and defect of T in $\mathcal{B}(X)$ are those of $\hat{T}$ in $\mathcal{B}(Ae)$. But $T \in \mathrm{Inv}(A)$ so $\hat{T} \in \mathrm{Inv}(\mathcal{B}(Ae))$. Suppose that $T \in \Phi(A)$ then $\mathrm{nul}(T) = 0 = \mathrm{def}(T)$ so, by F.2.8, $T \in \mathrm{Inv}(A)$, hence $T \in \mathrm{Inv}(A)$ modulo $\mathcal{K}(X)$.

However, the unital Banach algebra $A/\mathcal{K}(X)$ is generated by the element $T + \mathcal{K}(X)$ so $\sigma_{A/\mathcal{K}(X)}(T + \mathcal{K}(X))$ has connected complement ((14) 19.5). Further

$$\{\lambda : |\lambda| = 1\} = \sigma_{\mathcal{B}(X)/\mathcal{K}(X)}(T + \mathcal{K}(X)) \subset \sigma_{A/\mathcal{K}(X)}(T + \mathcal{K}(X)).$$

Therefore $\{\lambda : |\lambda| \leq 1\} \subset \sigma_{A/\mathcal{K}(X)}(T + \mathcal{K}(X))$ which contradicts the fact that $T \in \mathrm{Inv}(A)$ modulo $\mathcal{K}(X)$.

This exhibits a drawback of the representation

$$\pi : x \to \hat{x} : A \to \mathcal{B}(Ae)$$

for a general primitive Banach algebra. Further investigations into this case have been carried out by Alexander ((4) §5).

If, however, A is a primitive C*-algebra then the representation π is more useful. In the first place, as we see in §C*.4, Ae can be given the inner product

$$\langle x, y\rangle e \;=\; ey^*xe \;=\; y^*x \quad (x, y \in Ae),$$

under which it becomes a Hilbert space in the algebra norm. π is then a faithful irreducible *-representation which is therefore an isometry. Hence

$$\sigma_A(x) \;=\; \sigma_{\pi(A)}(\hat{x}) \;=\; \sigma_{\mathcal{B}(Ae)}(\hat{x}) \quad (\text{BA.4.2}).$$

Further, the converse of F.2.6, is valid in this case. In fact an examination of C*.4.2 and C*.4.3 shows that, since (0) is the only primitive ideal of A which does not contain soc(A) by BA.3.5, Λ in C*.4.3 becomes a singleton set, the π_2 in C*.4.3 is dispensable and we can take π to the representation defined above. Thus we have

F.4.3 THEOREM. *Let* A *be a primitive unital C*-algebra with* $e \in \text{Min}(A)$, *then*

(i) $\pi(\text{soc}(A)) = \mathcal{F}(Ae)$;
(ii) $\pi(\overline{\text{soc}(A)}) = \mathcal{K}(Ae)$;
(iii) $\pi(\mathcal{R}(A)) = \mathcal{R}(Ae) \cap \pi(A)$;
(iv) $\pi(\Phi(A)) = \Phi(Ae) \cap \pi(A)$.

($\mathcal{R}(A)$ is the set of Riesz elements of A relative to soc(A) defined in R.1.1).

F.2.3 and F.2.4 contain a definition of rank for elements of a primitive Banach algebra as well as a characterisation of the socle as the set of elements of finite rank. An alternative definition of finite rank elements via the wedge operator is given in C*.1.1 (x is of finite rank in A if $x \wedge x \in \mathcal{F}(A)$) and we show that, in a C*-algebra, the set of finite rank elements is equal to the socle (C*.1.2). Alexander ((4) 7.2) has extended this result to semisimple algebras. In primitive algebras the two definitions are equivalent.

Returning to Fredholm theory, Barnes' ideas for semisimple algebras were extended by Smyth (83) to general Banach algebras and this approach is followed here in §F.3. Pursuing suggestions of Barnes (8) and Pearlman (69)

Smyth introduced the index function (F.3.5) to cope with the problem that, if $x \in \Phi(A)$ and $\mathrm{ind}(x) = 0$, then x is not always decomposable into the sum of an invertible plus an inessential element. The original example of this in F.3.10 is due to Pearlman (69). Further information on this decomposition in the operator case is given by Murphy and West (62) and Laffey and West (55). Let X be a Banach space with $T \in \Phi(X)$, then it is shown that $T = V + F$ where V is left (right) invertible according as $i(T) \leq 0$ (≥ 0), $F \in \mathcal{F}(X)$ and the decomposition may be chosen so that $[V,F]^2 = 0$ where $[V, F] = VF - FV$. This result is best possible, in that it is not always possible to choose a decomposition such that $[V, F] = 0$, for example, if $i(T) = 0$, $T = V + F$ where $V \in \mathrm{Inv}(\mathcal{B}(X))$, $F \in \mathcal{F}(X)$ and $[V, F] = 0$ then, either $T \in \mathrm{Inv}(\mathcal{B}(X))$, or zero is a pole of finite rank of T. Using the techniques of this chapter these results can also be transplanted into Banach algebras. The index function for Fredholm elements in a general Banach algebra has also been defined by Kraljević, Suljagić and Veselić (110) making use of the concept of degenerate elements discussed in §R.5.

If A is a non-unital algebra then, in order to carry out Fredholm theory, one may adjoin a unit and proceed as in this chapter. (This will be necessary in Chapter R, for Riesz theory must be done in a non-unital setting). However as Barnes (8) and Smyth (83) showed, a different approach may be adopted. We say that $x \in A$ is *quasi-invertible* modulo an ideal F if there exists $y \in A$ such that $x + y - xy$, $x + y - yx \in F$. The set $\Psi = \Psi_F(A)$ is the set of all such elements. Let $I = k(h(F))$ and let $\mathcal{R}$ denote the set of elements in A all of whose scalar multiples lie in Ψ. Then $\mathcal{R}$ and I are the set of Riesz and inessential elements of A (respectively) relative to F. The elements of Ψ are called the *quasi-Fredholm* elements of A relative to F. We confine ourselves here to stating some useful results in quasi-Fredholm theory. The first follows from the fact that a quasi-invertible idempotent must be zero.

F.4.4 THEOREM. *Every idempotent of* Ψ *lies in* F.

In operator theory much interest has been focussed implicitly upon the quasi-Fredholm ideals including the ideals of finite rank, compact, strictly singular and inessential operators. In the algebraic context we note the following very general result starting with any quasi-Fredholm ideal J.

The proof depends on elementary properties of the radical and the fact that we can identify the structure space of A/J with the hull of J ((83) 4.2).

F.4.5 THEOREM. *Let* J *be an ideal of the algebra* A *such that* $F \subset J \subset \Psi$, *then*

(i) $x \in \Psi \Leftrightarrow x + J$ *is quasi-invertible in* A/J;

(ii) $x \in I \Leftrightarrow x + J$ *is in the radical of* A/J;

(iii) $x \in R \Leftrightarrow x + J$ *is quasinilpotent in* A/J;

(iv) $F \subset J \subset I \subset R \subset \Psi$;

(v) $h(F) = h(J) = h(I)$;

(vi) I *is the largest left or right ideal lying in* Ψ;

(vii) A/J *is semisimple* $\Leftrightarrow J = I$.

F.4.6 COROLLARY. *Any one of the sets* I, R, Ψ, $h(F)$ *uniquely determines each of the others.* ((83) 4.3)

The results thus far are valid for an arbitrary ideal F of A, for index theory we need to restrict F to lie in $\mathrm{psoc}(A)$.

The monograph (71), §A gives an interesting account of Fredholm theory for linear operators on linear spaces with no reference to topology. Related work is due to Kroh (54). Mizori-Oblak (59) studies elements of a Banach algebra whose left regular representations are Fredholm operators. If one is concentrating on an ideal F the choice of the ideal is important. This is exhibited by Yang (98) who studies operators on a Banach space invertible modulo the closed ideal of weakly compact operators. If the space is reflexive then every bounded linear operator is weakly compact and the Fredholm theory becomes trivial.

There have been many extensions of the classical Fredholm theory of linear operators of which the most important is the theory of semi-Fredholm operators.

F.4.7 DEFINITION. $T \in B(X)$ is *semi-Fredholm* if $T(X)$ is closed in X and if either $n(T)$ or $d(T)$ is finite.

The basic results for semi-Fredholm operators are given in (25), this class of operators has proved of central importance in modern spectral theory.

F.4.8 DEFINITION. An element x in a semisimple unital algebra A is *semi-Fredholm* if it is either left or right invertible modulo $\mathrm{soc}(A)$.

By F.1.10 if x is left (right) invertible modulo soc(A), x has right (left) Barnes idempotents in soc(A) so we could use the methods of this chapter to develop a semi-Fredholm theory in Banach algebras. In this monograph we confine ourselves to Fredholm theory.

Another extension of the classical theory leads to operators which have generalised inverses, or generalised Fredholm operators (named *relatively regular* operators by Atkinson (5)).

F.4.9 DEFINITION. (i) $T \in \mathcal{B}(X)$ has a *generalised inverse* $S \in \mathcal{B}(X)$ if $TST = T$, $STS = S$.
(ii) $T \in \mathcal{B}(X)$ is a *generalised Fredholm* operator if $T(X)$ is closed in X and both $\ker(T)$ and $T(X)$ are complemented subspaces in X.

F.4.10 THEOREM. $T \in \mathcal{B}(X)$ *has a generalised inverse* $\Longleftrightarrow$ T *is a generalised Fredholm operator.*

Proof. Let T have a generalised inverse S, then if $E = TS$, $F = ST$, it follows that $E = E^2$, $F = F^2$ and $I - E$, $I - F$ are left and right Barnes idempotents for T in $\mathcal{B}(X)$. We collect the following information.

$TS = E \Rightarrow E(X) \subset T(X)$, $\ker(S) \subset \ker(E)$;

$ST = F \Rightarrow F(X) \subset S(X)$, $\ker(T) \subset \ker(F)$;

$TF = T \Rightarrow \ker(F) \subset \ker(T)$;

$ET = T \Rightarrow T(X) \subset E(X)$;

$SE = S \Rightarrow \ker(E) \subset \ker(S)$;

$FS = S \Rightarrow S(X) \subset F(X)$.

Collating these results we see that

$$T(X) = E(X), \quad \ker(T) = \ker(F), \text{ and } \quad S(X) = F(X), \quad \ker(S) = \ker(E),$$

so both S and T are generalised Fredholm operators.

Conversely, let T be a generalised Fredholm operator, then the pictorial part of the proof of Atkinson's theorem (0.2.2) shows how to construct a generalised inverse S and it follows at once that $TST = T$, $STS = S$ ●

Generalised Fredholm theory for operators has been studied by Caradus (22), (23), (24), Yang (97), Treese and Kelly (90), among others. The class of generalised Fredholm operators on a Banach space contains all the projections in $\mathcal{B}(X)$ so one cannot expect such a tightly organised theory as in the classical Fredholm case, for example, this class is not, in general, open, or closed under compact perturbations, but we do have results of the following type ((22) Corollary 1).

F.4.11 THEOREM. *Let* T *be a generalised Fredholm operator on* X *and let* $V \in \mathcal{B}(X)$ *satisfy* $||V|| < ||S||^{-1}$, *where* S *is a generalised inverse of* T *and either* $\ker(V) \supset \ker(T)$ *or* $V(X) \subset T(X)$, *then* $T - V$ *is a generalised Fredholm operator.*

If T is generalised Fredholm its generalised inverse is not unique but, in Hilbert space, there exists a unique generalised inverse S such that the projections E and F are hermitean. Such an inverse is called a Moore-Penrose inverse in the matrix case (of course every matrix has a Moore-Penrose inverse), and this concept has recently proved to have many important applications. (A bibliography with 1700 items is contained in (64)). This situation has been algebraicised as follows: a semigroup S is called an *inverse semigroup* if each element $x \in S$ has a unique inverse y such that $xyx = x$, $yxy = y$. The structure of these semigroups is somewhat tractable and they have been objects of considerable study.

The Fredholm theory which we have developed in this monograph has as its outstanding characteristic an intimate connection with spectral theory. It has little connection with the Fredholm theory of Breuer (18), (19) extended by Olsen (68), based on the concept of a dimension function in von-Neumann algebras ((25) Chapter 6). Harte (106) has investigated Fredholm theory relative to a general Banach algebra homomorphism.

Coburn and Lebow ((25) Chapter 6) define a *generalised index* on an open semigroup of a topological algebra to be any homomorphism to another semigroup which is constant on connected components of the first semigroup. Of course, our theory fits into this very general framework and by specialising a little we obtain results (due to G.J. Murphy) on the existence and uniqueness of an index defined in a Banach algebra.

Let A denote a unital Banach algebra with proper closed ideal K and let Φ denote the set of elements of A invertible modulo K. Then Φ is an open multiplicative semigroup, $\mathrm{Inv}(A) \subset \Phi$, and $\Phi + K \subset \Phi$.

F.4.12 DEFINITION. A continuous semigroup homomorphism $i : \Phi \to G$ onto a discrete group G with unit element e is an *index* if, for $x \in \Phi$, $i(x) = e \Longleftrightarrow x \in \mathrm{Inv}(A) + K$.

It follows at once from the definition that $i(x + z) = i(x)$ $(x \in \Phi, z \in K)$, and that if $x \in \Phi$, there exists $\varepsilon > 0$ such that $y \in \Phi$ and $||x - y|| < \varepsilon \Rightarrow i(x) = i(y)$.

Our uniqueness result is somewhat surprising, roughly it states that, for a fixed K, the index is unique. To make this precise we need

F.4.13 DEFINITION. If $i : \Phi \to G$ and $j : \Phi \to H$ are indices they are *equivalent* if there is a group isomorphism $\theta : G \to H$ such that the following diagram commutes

$$\begin{array}{ccc} \Phi & \xrightarrow{i} & G \\ & \searrow_{j} & \downarrow \theta \\ & & H \end{array}$$

F.4.14 THEOREM. *There is, at most, one index up to equivalence.*

<u>Proof</u>. Let $x, y \in \Phi$ be such that $i(x) = i(y)$. Now there exists $u \in \Phi$ such that $uy - 1, yu - 1 \in K$. Clearly $i(y)^{-1} = i(u)$ so $e = i(x)i(u) = i(xu)$. Thus $xu = w + k$ for some $w \in \mathrm{Inv}(A)$ and $k \in K$. Multiply on the right by y to get $x = wy + k'$ where $k' \in K$. Thus $j(x) = j(w)j(y) = j(y)$, since $j(w) = e$. Now we can define a map $\theta : G \to H$ by $(\theta o i)(x) = j(x)$, and it follows immediately that θ is an isomorphism ●

Let $\psi : A \to A/K$ be the canonical homomorphism. The existence theorem is as follows.

F.4.15 THEOREM. (i) *An index exists* $\Longleftrightarrow \psi(\mathrm{Inv}(A))$ *is a closed normal subgroup of* $\mathrm{Inv}(A/K)$.

(ii) *If the condition in* (i) *is satisfied the group* $G = \mathrm{Inv}(A/K)/\psi(\mathrm{Inv}(A))$ *is discrete, and an index may be defined by setting*

$$i(x) = \psi(x)\psi(\mathrm{Inv}(A)) \qquad (x \in \Phi).$$

<u>Proof</u>. (i) Suppose that $j : \Phi \to H$, then the map $\theta : Inv(A/K) \to H : \psi(x) \to j(x)$ is a well defined group homomorphism onto H, with $\ker(\theta) = \psi(Inv(A))$. Since ψ is open, θ is continuous, thus $\psi(Inv(A))$ is a closed normal subgroup of $Inv(A/K)$.

Conversely, suppose $\psi(Inv(A))$ is a closed normal subgroup of $Inv(A/K)$. $Inv(A)$ is open in A, so $\psi(Inv(A))$ is open in A/K. Hence $G = Inv(A/K)/\psi(Inv(A))$ is a discrete group.

Part (ii) now follows easily ●

The abstract index defined here is not suitable for spectral theory, for example, there is no possibility of obtaining an analogue of the punctured neighbourhood theorem (0.2.7). In a sense, as the next result shows, any Fredholm index which gives rise to a satisfactory spectral theory is encompassed within the work of this chapter. As we have seen, if K is an inessential ideal, then the results of the classical spectral theory of bounded linear operators extend to Banach algebras. Now we show (informally) that if the results of classical Fredholm theory extend, then K must be an inessential ideal.

We shall make use of the characterisation of inessential ideals in R.2.6 as those ideals in which each element has zero as the only possible accumulation point of its spectrum. Suppose that i is a generalised index and that, relative to the ideal K, the results of classical Fredholm theory are valid. Let $x \in K$ and $0 \neq \lambda \in \sigma(x)$. We need to show that λ is an isolated point of $\sigma(x)$. It is clearly sufficient to do so for each $\lambda \in \partial\sigma(x)$. Since $x \in K$, $\lambda - x$ is invertible modulo K, hence $\lambda - x \in \Phi$. Thus, by the punctured neighbourhood theorem, there exists $\varepsilon > 0$ such that for $0 < |\mu - \lambda| < \varepsilon$, $\nu(\mu - x)$ and $\delta(\mu - x)$ are constant. But this punctured neighbourhood contains points of $\rho(x)$ hence $\nu(\mu - x)$ and $\delta(\mu - x)$ are both zero for $0 < |\mu - \lambda| < \varepsilon$ It follows by the classical theory that this punctured neighbourhood lies in $\rho(x)$, hence λ is an isolated point of $\sigma(x)$ and K is therefore an inessential ideal.

An element of an algebra is *algebraic* if it satisfies a polynomial identity, while an algebra is *algebraic* if every element therein is algebraic. The *algebraic kernel* of an algebra is the maximal algebraic ideal of the algebra. Its existence is demonstrated in (48) p.246-7 where it is shown to contain every right or left algebraic ideal.

The original setting for algebraic Fredholm theory was a semisimple Banach algebra and it was in this context, and relative to the socle, that Barnes(7) developed the theory in 1968. In 1969 he extended it to semiprime algebras. In the general case the socle does not always exist and, for this reason, Smyth (83) and Veselić (93) independently developed Fredholm theory relative to the algebraic kernel. In fact Smyth has shown ((84)§3) that the algebraic kernel of a semisimple Banach algebra is equal to the socle. A little more effort extends this result to semiprime Banach algebras. If A is a general Banach algebra and if Smyth's result is applied to the quotient algebra A' = A/rad(A) it follows that the algebraic kernel of A is contained in the presocle.

R Riesz theory

In this chapter the Ruston characterisation of Riesz operators (0.3.5) is used to define Riesz elements of a Banach algebra relative to any closed two-sided proper ideal, and elementary algebraic properties of Riesz elements are developed in §R.1 in this general setting. It transpires, however, that in order to obtain the deeper spectral theory of Riesz elements the ideal must be an inessential ideal and such a situation is investigated in §R.2. Finally the theory of Riesz algebras is developed in §R.3 and examples of Riesz algebras are listed in §R.4. Note that the algebras considered in this chapter will not necessarily be unital.

R.1 Riesz elements: algebraic properties

Let A be a Banach algebra and let K be a proper closed ideal of A.

R.1.1 DEFINITION. $x \in A$ is a *Riesz element* of A (relative to K) if $r(x + K) = 0$. $\mathcal{R}_K(A) = \mathcal{R}(A) = \mathcal{R}$ (when K is unambiguous from the context) will denote the set of Riesz elements of A.

This definition is motivated by the Ruston characterisation of Riesz operators (0.3.5). In the next section, having restricted K to be an inessential ideal we shall demonstrate the familiar spectral properties of Riesz elements.

Let $[x,y] = xy - yx$ denote the commutator of x and y. We have the following analogues of 0.3.6 and 0.3.7.

R.1.2 THEOREM. (i) $x \in \mathcal{R}$, $y \in K \Rightarrow x + y \in \mathcal{R}$;

(ii) $x \in \mathcal{R}$, $y \in A$ *and* $[x,y] \in K \Rightarrow xy, yx \in \mathcal{R}$;

(iii) $x,y \in \mathcal{R}$ *and* $[x,y] \in K \Rightarrow x + y \in \mathcal{R}$;

(iv) $x_n \in \mathcal{R}$ $(n \geq 1)$, $x_n \to x$ *in* A *and* $[x_n,x] \in K$ $(n \geq 1) \Rightarrow x \in \mathcal{R}$.

Proof. Apply the basic properties of the spectral radius to elements in A/K ●

R.1.3 THEOREM. *Let* $x \in A$ *and* $f \in \mathrm{Hol}(\sigma(x))$, *then*

(i) $x \in \mathcal{R}$ *and* $f(0) = 0 \Rightarrow f(x) \in \mathcal{R}$;

(ii) $f(x) \in \mathcal{R}$ *and* f *does not vanish on* $\sigma(x)\setminus\{0\} \Rightarrow x \in \mathcal{R}$;

(iii) (*if* A *is unital*) $x \in \Phi_K(A)$ *and* f *does not vanish on* $\sigma(x)\setminus\{0\} \Rightarrow f(x) \in \Phi_K(A)$.

Proof. (i) is a consequence of R.1.2 (ii), observing that $f(0) = 0 \Rightarrow f(x) = xg(x)$ where $g \in \mathrm{Hol}(\sigma(x))$.

Using the Cauchy integral representation of $f(x)$ one immediately verifies that if $x \in A$ and $f \in \mathrm{Hol}(\sigma(x))$, then $f \in \mathrm{Hol}(\sigma(x + K))$ since $\sigma(x + K) \subset \sigma(x)$ and $f(x + K) = f(x) + K$.

(ii) Since $f(x) \in \mathcal{R}$,

$$\sigma(f(x) + K) = \sigma(f(x + K)) = f(\sigma(x + K)) = \{0\}.$$

Now $\sigma(x + K) \subset \sigma(x)$, so, by hypothesis, $\sigma(x + K) = \{0\}$, hence $x \in \mathcal{R}$.

(iii) $x \in \Phi_K(A) \Rightarrow 0 \notin \sigma(x + K)$. Now $\sigma(x + K) \subset \sigma(x)$, therefore f does not vanish on $\sigma(x + K)$, so

$$0 \notin f(\sigma(x + K)) = \sigma(f(x + K)) = \sigma(f(x) + K)),$$

thus $f(x) \in \Phi_K(A)$ ●

Next we give two characterisations of the radical of a unital Banach algebra which lead to characterisations of the kernel of the hull of K. The characterisation involving $\mathrm{Inv}(A)$ is well known (BA.2.8), while that involving the set of quasinilpotent elements $\mathcal{Q}(A)$ is due to Zemánek (104). We recall that if ψ is the canonical quotient homomorphism $A \to A/K$ then $\psi(k(h(K))) = \mathrm{rad}(A/K)$ (BA.2.3).

R.1.4 THEOREM. *Let* A *be a unital Banach algebra, then*
$\mathrm{rad}(A) = \{x \in A : x + \mathrm{Inv}(A) \subset \mathrm{Inv}(A)\} = \{x \in A : x + \mathcal{Q}(A) \subset \mathcal{Q}(A)\}$.

R.1.5 COROLLARY. *Let* A *be unital then*
$k(h(K)) = \{x \in A : x + \Phi_K(A) \subset \Phi_K(A)\} = \{x \in A : x + \mathcal{R} \subset \mathcal{R}\}$.

R.2 Riesz elements: spectral theory

Recall that if A is a Banach algebra then $A' = A/\mathrm{rad}(A)$ and the ideal $I(A)$ of inessential elements of A is defined by

$$I(A) = \bigcap\{P \in \Pi(A) : P' \supset soc(A')\}.$$

We, henceforth, insist that K is a closed inessential ideal of A, that is, that K is closed ideal of A and $K \subset I(A)$. Our Riesz theory will be carried out relative to this fixed ideal K, so we shall drop the subscript from Φ_K and R_K.

We are going to deduce the spectral properties of Riesz elements from the Fredholm theory of Chapter F wherein it is assumed that A is unital. Thus, from R.2.1 to R.2.6, when we use results from Chapter F, A will always be unital and, at the end of the section, we shall show how these results may be extended to non-unital algebras.

R.2.1 DEFINITION. Let A be a unital Banach algebra. If $x \in A$, a complex number λ is called a *Fredholm point* of x if $\lambda - x \in \Phi$. The *Fredholm* or *essential spectrum* of x in A is defined to be the set

$$\omega(x) = \{\lambda \in \mathbb{C} : \lambda \text{ is not a Fredholm point of } x\}.$$

The *Weyl spectrum* of x is defined to be the set

$$W(x) = \bigcap_{y \in K} \sigma(x + y).$$

The complex number λ is called a *Riesz point* of x if either $\lambda - x$ is invertible, or if λ is a Fredholm point of x which is an isolated point of $\sigma(x)$. The *Riesz spectrum* or *Browder essential spectrum* of x in A is defined to be the set

$$\beta(x) = \{\lambda \in \mathbb{C} : \lambda \text{ is not a Riesz point of } x\}.$$

We note that $\omega(x)$, $W(x)$ and $\beta(x)$ are all compact subsets of $\mathbb{C}$ and the inclusion

$$\omega(x) \subset W(x) \subset \beta(x) \subset \sigma(x),$$

is valid for $x \in A$ Obviously $\omega(x) = \sigma_{A/K}(x + K)$, whenever K is proper, and since $\sigma_{A/K}(x + K) \subset \sigma(x + y)$ $(y \in K)$, it follows that $\omega(x) \subset W(x)$. Clearly $W(x)$ is a compact set. The inclusion $W(x) \subset \beta(x)$ is a consequence of the next theorem R.2.2. Clearly, $\beta(x)$ is a closed subset of $\sigma(x)$.

R.2.2 THEOREM. *Let* A *be a unital Banach algebra and* $x \in A$, *then*

$$W(x) = \{\lambda \in \mathbb{C} : \lambda - x \text{ is not a Fredholm element of } A \text{ of index-function zero}\}.$$

<u>Proof</u>. Taking complements the result may be restated as follows,

$$\bigcup_{y \in K} \rho(x + y) = \{\lambda \in \mathbb{C} : \lambda - x \in \Phi(A) \text{ and } \iota(\lambda - x) = 0\},$$

and, by using F.3.11, as

$$\bigcup_{y \in K} \rho(x + y) = \{\lambda \in \mathbb{C} : \lambda - x \in \mathrm{Inv}(A) + K\}.$$

This last statement is true since

$$\lambda \in \bigcup_{y \in K} \rho(x + y) \iff \lambda - x - y \in \mathrm{Inv}(A) \text{ for some } y \in K,$$

$$\iff \lambda - x \in \mathrm{Inv}(A) + K \quad \bullet$$

R.2.3 LEMMA. *A Riesz point of* x *which is in* $\sigma(x)$ *is an isolated point of* $\sigma(x)$ *and* $p(\lambda,x) \in K$.

<u>Proof</u>. Let λ be a Riesz point of x which lies in $\sigma(x)$. By the definition λ is isolated in $\sigma(x)$, hence the associated spectral idempotent

$$p(\lambda,x) = \frac{1}{2\pi i}\int_{\Gamma}(z - x)^{-1}dz \in A,$$

where Γ is a circle in $\rho(x)$ surrounding λ but no other point of $\sigma(x)$. Since λ is a Fredholm point of x, $\lambda \in \rho_{A/K}(x + K)$, hence the associated spectral idempotent

$$p(\lambda, x + K) = p(\lambda, x) + K = 0 \text{ in } A/K.$$

Therefore $p(\lambda, x) \in K$ ●

R.2.4 THEOREM. *Let* A *be a unital Banach algebra and* $x \in A$. *Then every Fredholm point of* x *lying in* $\partial\sigma(x)$ *is a Riesz point of* x.

Proof. If $\lambda \in \partial\sigma(x)$ then every neighbourhood of λ contains points of $\rho(x)$. If, in addition, $\lambda - x \in \Phi$, there exists $\varepsilon > 0$ such that $\mu - x \in \Phi$ for $|\mu - \lambda| < \varepsilon$, and, by the punctured neighbourhood theorem (F.3.9), $\nu(\mu - x)$, $\delta(\mu - x)$ are constant for $0 < |\mu - \lambda| < \varepsilon$. It follows that

$$\nu(\mu - x) = 0 = \delta(\mu - x) \text{ for } 0 < |\mu - \lambda| < \varepsilon,$$

thus, by F.3.7, this punctured neighbourhood of λ lies in $\rho(x)$. Therefore λ is an isolated point of $\sigma(x)$ which is, by definition, a Riesz point ●

The next result corresponds to the Ruston characterisation of Riesz operators (O.3.5).

R.2.5 COROLLARY. (Ruston characterisation) *Let* A *be a unital Banach algebra and* $x \in A$. *Then* $x \in R \iff$ *each* $\lambda \in \sigma(x)\setminus\{0\}$ *is an isolated point of* $\sigma(x)$ *and* $p(\lambda, x) \in K$.

Proof. If $x \in R$, then $r(x + K) = 0$, so if $0 \neq \lambda \in \sigma(x)$, then $\lambda - x + K \in \text{Inv}(A/K)$, hence $\lambda - x \in \Phi$. If $0 \neq \lambda \in \partial\sigma(x)$, then, by R.2.4, λ is a Riesz point of $\sigma(x)$. Hence every non-zero boundary point of $\sigma(x)$ is isolated in $\sigma(x)$. It follows that $\sigma(x)\setminus\{0\}$ is a discrete set of Riesz points of x, and, by R.2.3, the associated spectral idempotents lie in K.

Conversely, if each non-zero point of $\sigma(x)$ is a Riesz point of x, the set $\{\lambda \in \sigma(x) : |\lambda| > \delta > 0\}$ is a finite set $\{\lambda_k\}_1^n$, say. The spectral idempotent associated with this set, $p = \sum_1^n p(\lambda_k, x) \in K$, and $r(x - px) < \delta$.

Thus $r(x + K) \leq r(x - px) < \delta$,

and since δ is arbitrarily small, $r(x + K) = 0$ ●

In terms of the Browder spectrum this result states that $x \in \mathcal{R} \iff \beta(x) \subset \{0\}$. Note that if X is a finite dimensional linear space and $T \in \mathcal{B}(X)$ then $\beta(T)$ is empty.

The next result characterises the ideals K relative to which we can carry out Riesz and Fredholm theory, and is important in the characterisation of Riesz algebras in §R.3. The theorem is valid for ideals which are neither closed nor two-sided.

R.2.6 THEOREM. *Let* A *be a unital Banach algebra and* J *a left or right ideal of* A. *Then* $J \subset I(A) \iff$ *zero is the only possible accumulation point of* $\sigma(x)$ *for each* $x \in J$.

Proof. If $J \subset I(A)$, then $x \in J \Rightarrow x \in I(A) \Rightarrow r(x + I(A)) = 0$, hence zero is the only accumulation point of $\sigma(x)$ by R.2.5.

Conversely, let J be a left ideal of A such that zero is the only possible accumulation point of $\sigma(x)$ $(x \in J)$. Since $\sigma_A(x) = \sigma_{A'}(x')$ we may, and do, assume that A is semisimple. Let $x \in J$ and $0 \neq \lambda \in \sigma(x)$. By hypothesis, λ is an isolated point of $\sigma(x)$, so the associated spectral projection $p = p(\lambda, x) \in A$. We show that $p \in J$. pAp is a Banach algebra with unit p and $\sigma_{pAp}(px) = \{\lambda\}$. Thus $px \in \mathrm{Inv}(pAp)$, hence there exists $u \in pAp$ such that $p = upx \in J$.

Now A is semisimple, hence $I(A) = k(h(\mathrm{soc}(A)))$, and our next task is to show that $p \in I(A)$. Suppose not, then $p \notin \mathrm{soc}(A)$. Clearly then $p = p_1 \notin \mathrm{Min}(A)$ so, by BA.3.10, there exists $x_1 \in p_1Ap_1$ such that $\sigma_{p_1Ap_1}(x_1)$ contains at least two points. The hypothesis implies that zero is the only possible accumulation point of $\sigma_{p_1Ap_1}(x_1)$, hence this set contains an isolated point $\mu \neq 0$. Set $q_2 = p(\mu, x_1) \in p_1Ap_1$, then $q_2p_1 = p_1q_2 = q_2 \neq p_1$ (so $q_2 \in J$), with similar relations holding for $q'_2 = p_1 - q_2$. Now $p_1 = q_2 + q'_2$ so, at least one of $q_2, q'_2 \notin \mathrm{soc}(A)$. Label this idempotent p_2, then $p_2 \in p_1Ap_1$, $p_2 \notin \mathrm{soc}(A)$, $p_1p_2 = p_2p_1 = p_2 \neq p_1$, and $p_2 \in J$.

Continuing this process we get an infinite strictly decreasing family $\{p_n\}_1^{\infty}$ of idempotents in J all lying outside $soc(A)$, and satisfying $p_n p_m = p_m p_n = p_n$ for $m > n$. Set $q_1 = p_1$, $q_n = p_n - p_{n-1}$ $(n \geq 2)$, then $\{q_n\}_1^{\infty}$ is a properly infinite orthogonal family of idempotents in J such that $q_n p = pq_n = q_n (n \geq 1)$. Now choose a properly infinite sequence $\{\lambda_n\}_1^{\infty}$ of distinct complex numbers such that $|\lambda_n| < 2^{-n}||q_n||^{-1}$ $(n \geq 1)$, and $\lambda_n \to 0$ as $n \to \infty$. Then $w = \sum_1^{\infty} \lambda_n q_n \in A$, and $w = wp \in J$. Now $p + w \in J$ and

$$1 + \lambda_n \in \sigma(p + q_n) \ (n \geq 1) \Rightarrow (1 + \lambda_n) \in \sigma(p + w) \ (n \geq 1).$$

Thus 1 is an accumulation point of $\sigma(p + w)$. This gives the required contradiction, so that $p(\lambda,x) \in I(A)$, if $\lambda \in \sigma(x)\setminus\{0\}$ and $x \in J$.

It now follows, as in the proof of R.2.5, that $r(x + I(A)) = 0$ $(x \in J)$. Therefore $\{x + I(A) : x \in J\}$ is a left ideal of the Banach algebra $A/I(A)$ consisting entirely of quasinilpotent elements. Thus

$$\{x + I(A) : x \in J\} \subset rad(A/I(A)) \quad (BA.2.8),$$

but $A/I(A)$ is semisimple since $I(A)$ is a kernel (BA.2.3), so $J \subset I(A)$ ●

Riesz elements relative to a closed ideal of a Banach algebra which is not necessarily unital have been defined in §R.1. In R.2.1 Riesz points of an element in a unital algebra have been defined relative to Fredholm points of the element. To extend this definition and the results of R.2.3, R.2.5 and R.2.6 so as to obtain a full Riesz theory in a non-unital algebra let A be non-unital and denote by A_1 the unitization of A. Then, for $\dim(A) = \infty$, $\sigma_{A_1}(x) = \sigma_A(x) \cup \{0\}$ if $x \in A$, and if J is an ideal of A, it is also an ideal of A_1. We consider the case of semisimple A (and A_1), the result for general A follows on factoring out the radical. Then, if $\dim(A) = \infty$, $I(A) = k(h(soc(A)))$ and it is clear that $soc(A) = soc(A_1)$. Now every primitive ideal of A is contained in a primitive ideal of A_1, also A is a primitive ideal of A_1, further if $P_1 \in \Pi(A_1)$ and $P_1 \not\supset A$ then $P_1 \cap A \in \Pi(A)$ ((48) p.206). It follows that $I(A) = I(A_1) \cap A$ and the proofs extend immediately. The case of finite dimensional A is trivial.

We shall need the following consequence of the punctured neighbourhood theorem in Chapter C*.

R.2.7 THEOREM. *Let* Ω *be a connected open set of Fredholm points of an element* x *in a semisimple unital Banach algebra* A. *If* $\mu - x \in \mathrm{Inv}(A)$ *for some* $\mu \in \Omega$, *then every point of* Ω *is a Riesz point of* x *and* $\sigma(x) \cap \Omega$ *is a countable discrete set.*

Proof. $\mathrm{ind}(\mu - x) = 0$, hence $\mathrm{ind}(\lambda - x) = 0$ $(\lambda \in \Omega)$ by continuity of the index. Also $\mathrm{nul}(\lambda - x) = 0 = \mathrm{def}(\lambda - x)$, except for a set of discontinuities D of the function $\lambda \to \mathrm{nul}(\lambda - x)$ $(\lambda \in \Omega)$, which must be countable by the punctured neighbourhood theorem. The result now follows from F.3.7 and R.2.4 ●

R.3 Riesz algebras: characterisation

No study of Riesz or Fredholm theory is complete without a detailed analysis of the inessential ideals introduced in §F.3. To this end we examine those algebras such that $A = I(A)$. Recall that $I(A) = k\{P \in \Pi(A) : \mathrm{soc}(A') \subset P'\}$ where A is an algebra over $\mathbb{C}$. We start with general algebraic considerations.

R.3.1 DEFINITION. An algebra A is a *Riesz algebra* if $A = I(A)$.

It follows immediately from the definition and the homeomorphism between the structure spaces of A and A' (BA.2.5), that A is a Riesz algebra $\Longleftrightarrow$ A' is a Riesz algebra; and that A is a semisimple Riesz algebra $\Longleftrightarrow$ $h(\mathrm{soc}(A))$ is empty $\Longleftrightarrow$ $A/\mathrm{soc}(A)$ is a radical algebra (BA.2.3), since, in this case $A = I(A) = k(h(\mathrm{soc}(A)))$. Thus, semisimple Riesz algebras are 'close' to their socles in the sense that the socle is contained in no primitive ideal. It follows, and will be illustrated in §R.4, that the class of Riesz algebras is a large one embracing many important special algebras.

The characterisation of Riesz Banach algebras is an immediate consequence of R.2.6.

R.3.2 THEOREM. (Smyth characterisation) *A Banach algebra* A *is a Riesz algebra* $\Longleftrightarrow$ *zero is the only possible accumulation point of* $\sigma(x)$ *for each* $x \in A$.

It follows that, if a Banach algebra A is a Riesz algebra, then $\sigma(x)\setminus\{0\}$ is a discrete set, $\sigma(x)$ is countable and $\partial\sigma(x) = \sigma(x)$ $(x \in A)$.

A simple consequence of this is

R.3.3 COROLLARY. *Let* B *be a closed subalgebra, and* J *a closed ideal, of the Banach algebra* A. A *is a Riesz algebra* $\Rightarrow$ B *and* A/J *are Riesz algebras.*

We have noted the discreteness of the non-zero spectrum for every element in a Riesz Banach algebra. For general Riesz algebras the structure space is discrete in the hull-kernel topology.

R.3.4 THEOREM. *If* A *is a Riesz algebra,* $\Pi(A)$ *is discrete.*

Proof. Without loss of generality take A to be semisimple. Then the set of accumulation points of $\Pi(A)$ is contained in h(soc(A)) (BA.3.6), and is therefore empty ●

The converse of R.3.4 is false, for the Calkin algebra of an infinite dimensional Hilbert space has zero as its only primitive ideal but it is not a Riesz algebra. However, we do get a converse result in the commutative case.

R.3.5 COROLLARY. *If* A *is a commutative Banach algebra then* A *is a Riesz algebra* $\Leftrightarrow$ $\Pi(A)$ *is discrete in the hull-kernel topology* $\Leftrightarrow$ $\Pi(A)$ *is discrete in the Gelfand topology.*

Proof. The equivalence of the discreteness of $\Pi(A)$ in the hull-kernel and Gelfand topologies follows from BA.3.7. The proof is completed by applying R.3.4, and, conversely, by noticing that if $\Pi(A)$ is discrete and A is semisimple then $h(soc(A)) = \phi$ (BA.3.8) ●

The interesting examples of Riesz algebras are non-unital as the next result indicates.

R.3.6 THEOREM. *A unital Banach algebra is a Riesz algebra* $\Leftrightarrow$ *it is finite dimensional modulo its radical.*

Proof. Take A to be semisimple and unital. If A is a Riesz algebra then, unless A = soc(A), it follows that A/soc(A) is a unital radical algebra which is impossible. Now if A = soc(A), $1 = e_1 + \ldots + e_n$ where $e_i \in Min(A)$ $(1 \leq i \leq n)$. Then $A = \sum_{i=1}^{n} \sum_{j=1}^{n} e_i A e_j$ and $\dim(e_i A e_j) \leq 1$ $(1 \leq i, j \leq n)$ (BA.3.3), hence A is finite dimensional. The converse is obvious ●

The next result is an immediate consequence of R.2.6 if A is a Banach algebra, but it is also true for general algebras (a proof of this is indicated in §5), so it is stated in full generality.

R.3.7 THEOREM. *A left (right) ideal* L *of an algebra* A *is a Riesz algebra* $\Longleftrightarrow$ $L \subset I(A)$.

An obvious corollary of this is that every left or right ideal of a Riesz algebra is itself a Riesz algebra.

R.4 Riesz algebras: examples

Let A be a Banach algebra and $x \in A$; the commutant of x, $Z(x) = \{y \in A : yx = xy\}$, is a closed subalgebra of A. Let $\tilde{x}$ denote the bounded linear map

$$\tilde{x} : y \to xy : Z(x) \to Z(x).$$

Then $\sigma_A(x) = \sigma_{\mathcal{B}(Z)}(\tilde{x})$, and we write the latter set $\sigma(\tilde{x})$. Recall that $\mathcal{R}(X)$ denotes the set of Riesz operators on the Banach space X. In R.4.1 and R.4.2 we shall be considering operators on the Banach spaces Z(x) and A.

R.4.1 COROLLARY. *If* A *is a Banach algebra such that* $\tilde{x} \in \mathcal{R}(Z(x))$ *for each* $x \in A$ *then* A *is a Riesz algebra.*

If $x \in A$ let L_x and R_x denote the operators of left and right multiplication by x on A, respectively.

R.4.2 COROLLARY. *A Banach algebra* A *is a Riesz algebra if any of the three following conditions hold,*

(i) $L_x \in \mathcal{R}(A)$ $(x \in A)$;

(ii) $R_x \in \mathcal{R}(A)$ $(x \in A)$;

(iii) $x_\wedge x : u \to xux \in \mathcal{R}(A)$ $(x \in A)$.

Proof. Let $x \in A$ and put $T = L_x$ or R_x or $x_\wedge x$. Clearly $Z(x) = Z$ is invariant under T so if (i), (ii) or (iii) hold then $T|Z \in \mathcal{R}(Z)$ ((23) 3.5.1). In cases (i) and (ii) this implies that $\tilde{x} \in \mathcal{R}(Z)$ and in case (iii) that $(\tilde{x})^2 = (x^2)^\sim \in \mathcal{R}(Z)$ which, by the Ruston characterisation (0.3.5), implies that $\tilde{x} \in \mathcal{R}(Z)$. Thus, in each case, by R.4.1, A is a Riesz algebra ●

R.4.3 EXAMPLE. The Banach algebra A is *compact* if $x_\wedge x$ is a compact operator on A $(x \in A)$. Compact Banach algebras have been studied by Alexander (4), and, by R.4.2, are Riesz algebras.

R.4.4 EXAMPLE. The Banach algebra A is *left (resp. right) completely continuous* L.C.C. (resp. R.C.C.) if L_x (resp. R_x) is a compact operator on A $(x \in A)$. These algebras have been studied by Kaplansky (50), and, by R.4.2, are Riesz algebras.

R.4.5 EXAMPLE. Let G be a locally compact abelian group. The Gelfand space of the commutative Banach algebra $L^1(G)$ is the dual group $\hat{G}$ in the induced topology. Now $\hat{G}$ is discrete $\Longleftrightarrow$ G is compact, so by R.3.5, $L^1(G)$ is a Riesz algebra $\Longleftrightarrow$ G is compact.

R.4.6 EXAMPLE. Let X be a Banach space and A a closed subalgebra of $\mathcal{B}(X)$ which is contained in $\mathcal{R}(X)$, then A is a Riesz algebra by R.3.2.

R.4.7 EXAMPLE. Let X be a Banach space, let $T \in \mathcal{R}(X)$ and let A be the uniformly closed subalgebra of $\mathcal{B}(X)$ generated by T, then $A \subset \mathcal{R}(X)$ and is, therefore, a Riesz algebra. To see this, note that if $p(z)$ is a polynomial vanishing at the origin then $p(T) \in \mathcal{R}(X)$ (0.3.7).

R.4.8 EXAMPLE. Let T be a non-nilpotent quasinilpotent operator on a Banach space and let $\mathcal{P}$ be the algebra generated by T and I (the algebra of polynomials in T). Then $\mathrm{Inv}(\mathcal{P})$ consists of the non-zero multiples of I, and I is the only non-zero idempotent in $\mathcal{P}$. Using a characterisation of the radical as in ((14) 24.17) we see that $\mathcal{P}$ is semisimple and since $\mathrm{Min}(\mathcal{P}) = \phi$, $\mathrm{soc}(\mathcal{P}) = (0)$. So $I(\mathcal{P}) = k(h(\mathrm{soc}(\mathcal{P}))) = (0)$, and $\mathcal{P}$ is not a Riesz algebra.

Now let A be the closure of $\mathcal{P}$ in $\mathcal{B}(X)$, then $A = \mathbb{C} \oplus \mathrm{rad}(A)$ so, by R.3.6, A is a Riesz algebra. Thus we have constructed a Riesz algebra with a dense subalgebra which is not a Riesz algebra.

R.5 Notes

Attempts to develop a theory of compact or weakly compact elements in a Banach algebra A pre-date the development of a Riesz theory. We list some of the early definitions.

M. Freundlich (34) 1949: t is a compact element of A if both operators $a \to ta$, $a \to at$ are compact on A.

T. Ogasawara (65) 1954: t is a weakly completely continuous element of A if both the above operators are weakly compact on A.

K. Vala (92) 1968: t is a compact element of A if the wedge operator $t_\wedge t : a \to tat$ is compact on A.

B. J. Tomiuk and P. K. Wong (77) 1972: t is a weakly semi-completely continuous element of A if the wedge operator is weakly compact on A.

These concepts have also been studied by Alexander (4), Bonsall (12), Kaplansky (50), Ogasawara and Yoshinaga (66), Ylinen (99), (101). None of the definitions is entirely satisfactory. This is illustrated in the case of the wedge operator of Vala by an example due to Smyth (86) of a semisimple Banach algebra A in which $x_\wedge x$ is compact ($x \in A$) but $x_\wedge y$ need not be. However, Ylinen (100) has shown that some of these definitions coincide with the reasonable definition of the set of compact elements in a C*-algebra given in C*.1.1. The following artificial definition has been proposed by Smyth for a general Banach algebra A. It has the merit that the compact elements form a closed two-sided ideal with many of the expected properties. Let $S = \{u \in A : u_\wedge u$ is a finite rank operator$\}$ and set $F = \{x \in A : x_\wedge u$ is of finite rank for each $u \in S\}$. (If A is semiprime then $F = S$). If $J = k(h(F))$ the compact elements are defined to be the set

$$\{x \in J : x_\wedge y \text{ is a compact operator for each } y \in F\}.$$

Riesz theory as presented here dates back to 1968 when Barnes (7) derived the spectral properties of inessential elements of a semisimple Banach algebra using the concept of ideals of finite order. His paper contains an analogue of the Ruston characterisation. A more general approach was adopted by Smyth (82), (84) who developed Riesz theory relative to a fixed ideal F of algebraic elements (those satisfying a non-trivial polynomial identity). He showed that in a semisimple Banach algebra the socle is the largest such ideal (termed the *algebraic kernel*) and obtained a Riesz theory (including the Ruston characterisation) for elements x such that $r(x + \bar{F}) = 0$. The algebraic kernel is the largest left or right ideal of algebraic elements and the proof of its existence in a complex algebra is due to Amitsur ((48)p.246-7).

Veselić (93) has demonstrated the existence of the algebraic kernel in a complex Banach algebra using function theoretic methods. He defines $x \in A$ to be *degenerate* if each element in Ax is algebraic. Thus Ax is a left ideal of algebraic elements of A, hence it lies in the algebraic kernel of A. Indeed, the set of degenerate elements of A is actually equal to the algebraic kernel and from this it follows that Veselić's set of compact elements is precisely our set of inessential elements. In Problem 8, Veselić asked if X is a Banach space and $T \in \mathcal{B}(X)$ is such that the algebra generated by ST is finite dimensional for each $S \in \mathcal{B}(X)$, is $S \in \mathcal{F}(X)$? The answer is yes, for ST is algebraic for each S, hence $\mathcal{B}(X)T$ is a left ideal of algebraic elements which is contained in the algebraic kernel which in turn is equal to the socle $\mathcal{F}(X)$. Puhl (72) has defined a trace functional in a suitable subalgebra of a Banach algebra.

We record two further results in Riesz theory, both due to Smyth ((83), 6.3, 6.4).

R.5.1 THEOREM. *The sets of Fredholm and Riesz points of an element in a commutative, unital Banach algebra coincide.*

Proof. Let A be a commutative, unital Banach algebra and let $x \in A$. Since $\rho(x)$ is a subset of both sets, we consider only the Fredholm and Riesz points in $\sigma(x)$.

If zero is a Riesz point of x in $\sigma(x)$ and if $p \in K$ is the associated spectral idempotent, then $x(1 - p) + p \in \text{Inv}(A)$, so $x + K \in \text{Inv}(A/K)$, hence zero is a Fredholm point of x.

Conversely, let zero be a Fredholm point of x in $\sigma(x)$, then, by R.2.4, it suffices to show that zero is an isolated point of $\sigma(x)$. Now there exists $y \in A$ such that $xy = 1 + k$, where $k \in K$. Since zero is the only possible accumulation point of $\sigma(k)$, then, either, $0 \in \rho(xy)$, or zero is an isolated point of $\sigma(xy)$. Now, by commutativity, $0 \in \rho(xy) \Rightarrow 0 \in \rho(x)$, which is false. Hence there exists a non-zero spectral idempotent p and

$$p = p(0, xy) = p(0, 1 + k) = p(-1, k) \in K.$$

Now $\sigma_{Ap}(xyp) = \{0\}$ and $\sigma_{A(1-p)}(xy(1-p)) = \sigma_A(x) \setminus \{0\}$.

Thus $0 \in \rho_{A(1-p)}(xy(1 - p))$ and again, by commutativity, $0 \in \rho_{A(1-p)}(x(1-p))$.

Now Ap is a Riesz algebra with unit p, hence, by R.3.6, it is finite dimensional modulo its radical. Thus $\sigma_{Ap}(xp)$ is a finite set which contains zero. So zero is isolated in $\sigma_A(x) = \sigma_{Ap}(x) \cup \sigma_{A(1-p)}(x(1-p))$ ●

The second result is a spectral mapping theorem for the Fredholm and Browder spectra.

R.5.2 THEOREM. *If* A *is a unital Banach algebra,* $x \in A$ *and* $f \in \mathrm{Hol}(\sigma(x))$ *then*

(i) $f(\omega(x)) = \omega(f(x))$;

(ii) $f(\beta(x)) = \beta(f(x))$.

Proof. (i) Since $\omega(x) = \sigma_{A/K}(x + K)$, this result is simply the spectral mapping theorem in A/K.

(ii) If $x \in A$, its bicommutant $Z_2 = Z_2(x)$ is the commutative Banach algebra

$$Z_2(x) = \{z \in A : [z,y] = 0 \text{ for each } y \text{ such that } [x,y] = 0\}.$$

Note that every spectral idempotent of x, and of $f(x)$, lies in Z_2. We consider Fredholm theory in the commutative Banach algebra in Z_2 relative to the ideal $K \cap Z_2$. (If $K \cap Z_2 = (0)$, the result reduces to the ordinary spectral mapping theorem). If $z \in K \cap Z_2$, then $\sigma_A(z) \setminus \{0\}$ is a discrete set, thus so is $\sigma_{Z_2}(z) \setminus \{0\}$. Hence $K \cap Z_2$ is a closed ideal of Z_2, and, by R.2.6, $K \cap Z_2 \subset I(Z_2)$. Now let ω_{Z_2} and β_{Z_2} denote the Fredholm and Browder spectra in Z_2. From the definition R.2.1, and by R.5.1,

$$\omega_{Z_2}(z) = \beta_{Z_2}(z) \quad (z \in Z_2).$$

Further, $\beta_A(x) = \beta_{Z_2}(x)$ since all the spectral idempotents of x lie in Z_2. The result now follows by applying (i) ●

Next we give the proof of R.3.7 in the setting of a general algebra.

Recall (F.3.1) that the presocle of an algebra A is defined to be the ideal $\mathrm{psoc}(A) = \{x \in A : x' \in \mathrm{soc}(A')\}$.

This ideal is of considerable interest in the study of Fredholm theory, it was introduced by Smyth and its elementary properties are outlined in (83). Clearly $I(A) = k(h(\text{psoc}(A)))$, and A is a Riesz algebra $\iff h(\text{psoc}(A)) = \phi \iff A/\text{psoc}(A)$ is a radical algebra. The first of the two lemmas required for the proof connects the set $\text{Min}(L/\text{rad}(L))$, where L is a left ideal of A, with the set $\text{Min}(A/\text{rad}(A))$.

R.5.3 LEMMA. *Let* L *be a left ideal of the algebra* A. *If* $x \in L$ *and* $x' \in \text{Min}(A')$ *then* $x + \text{rad}(L) \in \text{Min}(L/\text{rad}(L))$. *Conversely, every element of* $\text{Min}(L/\text{rad}(L))$ *is of the form* $x + \text{rad}(L)$ *for some* $x \in L$ *such that* $x' \in \text{Min}(A')$.

Proof. Suppose $x \in L$ and $x' \in \text{Min}(A')$. Since $x^2 - x \in L \cap \text{rad}(A) \subset \text{rad}(L)$, it follows that $x + \text{rad}(L)$ is an idempotent of $L/\text{rad}(L)$. Choose $u \in xLx\backslash\text{rad}(L)$, then $u \in xAx\backslash\text{rad}(A)$, hence, since $x'A'x'$ is a division algebra with unit x', there exists $v \in xAx \subset L$, such that $u'v' = x' = v'u'$. Write $w = xvx \in xLx$. Then $u'w' = x' = w'u'$, so $uw - x, wu - x \in \text{rad}(A) \cap L \subset \text{rad } L$. It follows that $xLx + \text{rad}(L)/\text{rad}(L)$ is a division algebra so $x + \text{rad}(L) \in \text{Min}(L/\text{rad }(L))$.

Conversely, suppose $y \in L$ is such that $y + \text{rad}(L) \in \text{Min}(L/\text{rad}(L))$. Write $x = y^4 \in L$ and note that $x + \text{rad}(L) = y + \text{rad}(L)$. We shall show that $x' \in \text{Min}(A')$. Let $t \in y^2Ay \cap \text{rad}(L)$, say $t = y^2zy$ for $z \in A$. Then since $y^2 - y \in \text{rad}(L)$, it follows that $yzy \in \text{rad}(L)$, and hence $At = (Ay)yzy \subset \text{rad}(L)$. Using a characterisation of the radical ((14) 24.16), it follows that $t \in \text{rad}(A)$, hence $y^2Ay \cap \text{rad}(L) \subset \text{rad}(A)$. Now $x^2 - x \in y^2Ay \cap \text{rad}(L) \subset \text{rad}(A)$, hence x' is an idempotent of A'. Choose $u \in xAx\backslash\text{rad}(A)$. Then $u \in yLy\backslash\text{rad}(L)$. Now since $yLy/\text{rad}(L)$ is a division algebra and $x = y$ modulo L, it follows that there exists $v \in A$ such that $uv - x, vu - x \in \text{rad}(L)$. Write $w = xvx$. Then $uw - x, wu - x \in y^2Ay \cap \text{rad}(L) \subset \text{rad}(A)$. It follows that $x'A'x'$ is a division algebra, hence $x' \in \text{Min}(A')$ ●

The second lemma connects the presocles of L and A.

R.5.4 LEMMA. *If* L *is a left ideal of the algebra* A *then* $\text{psoc}(A) \cap L \subset \text{psoc}(L)$.

Proof. Choose $x \in \mathrm{psoc}(A) \cap L$. Then $x' \in \mathrm{soc}(A')$ and, because the socle is the sum of the minimal left ideals, there exist $\{x'_k\}_1^n \subset A'x' \cap \mathrm{Min}(A')$ such that $x' = \sum_1^n x'x'_k$. Clearly we may assume that $\{x_k\}_1^n \subset Ax \subset L$. Then, by R.5.3, $\{x_k + \mathrm{rad}(L)\}_1^n \subset \mathrm{Min}(L/\mathrm{rad}(L))$, and

$$x - \sum_1^n xx_k \in \mathrm{rad}(A) \cap L \subset \mathrm{rad}(L).$$

Therefore $x + \mathrm{rad}(L) \in \mathrm{soc}(L/\mathrm{rad}(L))$, hence $x \in \mathrm{psoc}(L)$ as required ●

We may now complete the proof of R.3.7. Suppose first that $L \subset I(A) = k(h(\mathrm{psoc}(A)))$. Consider the standard isomorphism

$$L/(L \cap \mathrm{psoc}(A)) \approx (L + \mathrm{psoc}(A))/\mathrm{psoc}(A).$$

The algebra on the right of this isomorphism is a left ideal of $I(A)/\mathrm{psoc}(A)$ which is a radical algebra, hence it also is a radical algebra, therefore so also is the algebra on the left. Then, by R.5.4, $L/\mathrm{psoc}(L)$ is a radical algebra, so L is a Riesz algebra.

Conversely, suppose that L is a Riesz algebra. Let P be a primitive ideal of L which contains $\mathrm{psoc}(A) \cap L$. Then, by R.5.3, every minimal idempotent of $L/\mathrm{rad}(L)$ lies in $P/\mathrm{rad}(L)$ thus, the ideal generated by these idempotents, $\mathrm{soc}(L/\mathrm{rad}(L)) \subset P/\mathrm{rad}(L)$. However, by the homeomorphism of structure spaces, $P/\mathrm{rad}(L)$ is a primitive ideal of $L/\mathrm{rad}(L)$ which therefore cannot contain $\mathrm{soc}(L/\mathrm{rad}(L))$, since L is a Riesz algebra. Thus no primitive ideal of L can contain $\mathrm{psoc}(A) \cap L$. It follows that $L/(L \cap \mathrm{psoc}(A))$ is a radical algebra, hence by the above isomorphism $(L + \mathrm{psoc}(A))/\mathrm{psoc}(A)$ is a radical left ideal of $A/\mathrm{psoc}(A)$ which, therefore, lies in the radical of $A/\mathrm{psoc}(A)$, which is $I(A)/\mathrm{psoc}(A)$. Therefore $L \subset I(A)$ ●

We conclude with some remarks on Riesz algebras. The theory presented here in §R.3 is due to Smyth (85). Riesz algebras are related to the class of modular annihilator algebras defined by Yood (103). (An algebra is *semiprime* if it has no non-zero ideals with square zero). A semiprime algebra is defined to be *modular annihilator* if every modular maximal left (right) ideal has a non-zero right (left) annihilator. If we restrict to the semisimple case, the classes of Riesz and modular annihilator algebras coincide. A summary of the properties and a list of examples of modular annihilator algebras is given by Barnes (9). The more important examples in this class (with the additional assumption of semisimplicity) are H^*-algebras (Ambrose (1)), dual algebras (Kaplansky (50), (51)), annihilator algebras (Bonsall and Goldie (15)) and Banach algebras with a dense socle. This class includes most of the important Banach algebras possessing minimal ideals.

Although a semiprime modular annihilator algebra is a Riesz algebra, a semiprime Riesz algebra need not be a modular annihilator algebra. For example if B is a radical Banach algebra with $\text{ran}(B) = (0)$ let A be $B \oplus \mathbb{C}1$. Then B is a maximal modular left ideal of A, but $\text{ran}(B)$ in A is $\{0\}$. Then A is not a modular annihilator algebra, but, by R.3.6, A is a Riesz algebra. Such a B can be constructed as the closure in $\mathcal{B}(X)$ of the algebra generated by a non-nilpotent quasinilpotent operator.

Observe that the result of R.4.5 holds also for compact non-abelian G. In fact, if dy denotes a fixed Haar measure on G, then, if $f, g \in L^1(G)$

$$(f*g)(x) = \int_G f(xy^{-1})g(y)dy$$

((45) 20.10). Since G is compact it is easy to verify that the operator $L_f : g \to f*g \quad (g \in L^1(G))$ is compact for each $f \in L^1(G)$. Hence $L^1(G)$ is LCC.

Kraljević and Veselić (109) define *spectrally finite* Banach algebras as those for which the spectrum of every element is a finite set. In fact these algebras are precisely those which are equal to their presocles. (109) contains an approach to the dimension concept in Banach algebras somewhat different from that of Barnes (8), (9).

C* C*-algebras

The wedge operator which has been defined on the algebra $\mathcal{B}(X)$ in Chapter 0 proves to be particularly effective in C*-algebras. This operator is studied in §1. §2 contains the decomposition theorems of West and Stampfli for operators in Hilbert space and their analogues in C*-algebras. A modification of a conjecture of Pelczyński leads, in §3, to a characterisation of Riesz algebras among C*-algebras. A representation is constructed in §4 which provides an isometric map of finite rank, compact, and Riesz elements of a C*-algebra onto the corresponding operators in Hilbert space. In §5 short proofs are provided for the range inclusion results of §0.4, which are available in Hilbert space.

C*.1 The wedge operator

The wedge operator defined in §0.6 has been used by Vala (91), (92), Alexander (4), and Ylinen (99) to define classes of finite rank and compact elements in Banach algebras, (in C*-algebras, Definition C*.1.1). In the case of the algebra of bounded linear operators on a Banach space these definitions are justified by Theorem 0.6.1, which shows that the finite rank and compact elements of this algebra are, precisely, the finite rank and compact operators on the Banach space. The definitions are unsatisfactory in a general Banach algebra; an example due to Vala (92) showing that the set of compact elements may fail to be closed under addition. The definitions are satisfactory, however, in the case of a C*-algebra, and in this section we demonstrate this fact.

C*.1.1 DEFINITION. An element x of a C*-algebra A is said to be *finite rank* or *compact* in A if the wedge operator $x \wedge x : a \to xax$ is a finite rank or compact operator on A respectively.

C*.1.2 THEOREM. soc(A) *is the set of finite rank elements in a C*-algebra* A.

Proof. Let $x \in soc(A)$, then $x = \sum_{1}^{k} a_i e_i$, where $a_i \in A$, $e_i \in Min(A)$ $1 \leq i \leq k$. Then

$$xAx = \left(\sum_{1}^{k} a_i e_i\right) A \left(\sum_{1}^{k} a_j e_j\right) = \sum_{i,j=1}^{k} a_i e_i A a_j e_j,$$

$$\subset \sum_{i,j=1}^{k} a_i e_i A e_j,$$

$$= \sum_{i,j=1}^{k} a_i \mathbb{C} b_{ij},$$

where each b_{ij} is a fixed element of $e_i A e_j$ (BA.3.3). Thus $\dim(xAx) < \infty$.

Conversely, suppose that $\dim(xAx) < \infty$. Now $xAx^*x \subset xAx$, so $\dim(xAx^*x) < \infty$, therefore $\dim(x^*xAx^*x) < \infty$. Since $x \in soc(A) \iff xx^* \in soc(A)$ (BA.4.4), it suffices to assume that x be self-adjoint. Let Z be the commutant of x in A. Now $\sigma(x \wedge x|Z) = \sigma_Z(x^2)$ is a finite set, thus so is $\sigma_A(x)$. Then $x = \sum_{1}^{k} \lambda_n e_n$, where $\lambda_n \neq 0$, $e_n = e^*_n = e^2_n$ $(1 \leq n \leq k)$, and $e_m e_n = 0$ $(m \neq n)$. Now, for each n,

$$\infty > \dim(xAx) \geq \dim(e_n xAxe_n) = \dim(e_n Ae_n).$$

From this it is easy to conclude that each e_n (and consequently x) $\in soc(A)$. For suppose that $\dim(eAe) < \infty$, where $e^2 = e \in A$, then eAe is a Riesz algebra which, by the Smyth characterisation (R.3.2), is equal to its own socle. But if $p \in Min(eAe)$ and $x \in A$

$$pxp = pexep = \lambda p$$

for some scalar λ, hence $p \in Min(A)$, so $eAe = soc(eAe) \subset soc(A)$ ●

C*.1.3 THEOREM. $\overline{soc(A)}$ *is the set of compact elements in a C*-algebra* A.

Proof. Let $x \in \overline{soc(A)}$. Choose $x_n \in soc(A)$ such that $||x_n|| \leq ||x|| + 1$, and $||x_n - x|| \to 0$. Then

$$||x_n \wedge x_n - x \wedge x|| \leq (2||x|| + 1)||x_n - x|| \to 0.$$

By C*.1.2, $x \wedge x$ is the uniform limit of operators of finite rank on A. Hence $x \wedge x$ is a compact operator on A.

Conversely, let x be a compact element of A, then $x \wedge x$ is a compact operator on A. Now the operator $x^*x \wedge x^*x$ being the composition of the operators

$$a \to ax^* \to xax^*x \to x^*xax^*x \qquad (a \in A)$$

is compact on A. Thus, since $x \in \overline{soc(A)} \Longleftrightarrow x^*x \in \overline{soc(A)}$ (BA.4.4), it is sufficient to assume that x is self-adjoint.

By considering the commutant of x and arguing as in C*.1.2, we see that $\sigma(x)$ has zero as its only possible accumulation point. Thus, if $\varepsilon > 0$, the set

$$\sigma(x) \cap \{\lambda: |\lambda| \geq \varepsilon\}$$

is finite $\{\lambda_1, \ldots, \lambda_n\}$, say. If e_i is the eigenprojection of x corresponding to λ_i then

$$e_i x = xe_i = \lambda_i e_i \quad (1 \leq i \leq n),$$

so $\lambda_i^{-2}(e_i \wedge e_i)(x \wedge x) = e_i \wedge e_i$,

hence $e_i \wedge e_i$ is a compact projection which is therefore of finite rank. By C*.1.2, $e_i \in soc(A)$ $(1 \leq i \leq n)$, hence $p = \sum_1^n e_i \in soc(A)$. Using self-adjointness

$$||x(1 - p)|| = r(x(1 - p)) < \varepsilon,$$

hence $x \in \overline{soc(A)}$ ●

C*.1.4 THEOREM. *The following statements are equivalent in a C*-algebra* A:

(i) $x \wedge x$ *is a Riesz operator on* A;

(ii) x *is a Riesz element of* A *relative to the ideal* $\overline{soc(A)}$;

(iii) $r(x + \overline{soc(A)}) = 0$.

Proof. (ii) and (iii) are equivalent by the definition, while the equivalence of (i) and (ii) follows exactly as in 0.6.1 ●

C*.2 Decomposition theorems

First we state our two decomposition theorems in a Hilbert space H, the West decomposition of a Riesz operator into the sum of a compact operator plus a quasinilpotent and Stampfli's generalisation of it using the Weyl spectrum. Then using the machinery already constructed these results are proved in a C*-algebra setting.

C*.2.1 THEOREM. (West decomposition) $T \in \mathcal{R}(H) \Rightarrow T = K + Q$ *where* $K \in \mathcal{K}(H)$, K *is normal*, $\sigma(T) = \sigma(K)$, *and* $Q \in \mathcal{Q}(H)$.

((25) 3.5).

Recall that, if T is a Riesz operator, every non-zero point of $\sigma(T)$ is index-zero Fredholm and that the Weyl spectrum of T is defined by

$$W(T) = \{\lambda \in \sigma(T) : \lambda - T \notin \Phi^{0}(H)\}.$$

Thus if $T \in \mathcal{R}(H)$, $W(T) \subset \{0\}$. Stampfli (88) has generalised C*.2.1 as follows,

C*.2.2 THEOREM. (Stampfli decomposition) $T \in \mathcal{B}(H) \Rightarrow T = K + S$ *where* $K \in \mathcal{K}(H)$, *and* $\sigma(S) = W(T)$.

((88) Theorem 4).

The extensions of these two results appear as corollaries of the next theorem. A is a unital C*-algebra and K a closed ideal of A.

C*.2.3 THEOREM. *Let* $x \in A$ *and suppose that* ω *is a subset of* $\sigma(x)$ *such that every point of* ω *is isolated in* $\sigma(x)$, *and that the corresponding spectral idempotent lies in* K, *then there exists a normal element* $y \in K$ *such that* $\sigma(x + y) = \sigma(x)\setminus\omega$.

Proof. We shall make use of the Gelfand-Naimark embedding of A into $\mathcal{B}(H)$ for a suitable Hilbert space H ((14) 38.10).

If ω is a finite set the proof is trivial, thus suppose not. Because the points of ω are isolated in the compact set $\sigma(x)$, ω must be a countable set $\{\lambda_k\}_1^\infty$, where the λ_k's are labelled in decreasing order of nearness to $\sigma(x)\setminus\omega$. Thus we may choose a countable set $\{\alpha_k\}_1^\infty \subset \sigma(x)\setminus\omega$, such that $\lambda_k - \alpha_k \to 0$. Let $p_k \in K$ be the spectral idempotent for x corresponding to λ_k, and put $s_n = \sum_1^n p_k$. Let q_n be the self-adjoint idempotent in K such that $q_n(H) = s_n(H)$ (BA.4.3). Now setting $q_o = 0$,

$$y_n = \sum_1^n (\alpha_k - \lambda_k)(q_k - q_{k-1}) \in K,$$

and since $\{q_k - q_{k-1}\}_1^\infty$ is a set of disjoint self-adjoint idempotents, y_n is normal for each n, hence $y_n \to y \in K$, and y is normal. It remains to show that $\sigma(x + y) = \sigma(x)\setminus\omega$.

Write $f_n = (x + y_n)|s_n(H)$ and $g_n = (x + \alpha_n - \lambda_n)|p_n(H)$. Then $f_1 = g_1$ and relative to the decomposition $s_n(H) = s_{n-1}(H) \oplus p_n(H)$ we have

$$f_n = \begin{bmatrix} f_{n-1} & * \\ 0 & g_n \end{bmatrix} \quad (n \geq 2).$$

Since $\sigma(g_n) = \{\alpha_n\}$, it follows, from (BA.4.5), that $\sigma(f_n) = \sigma(f_{n-1}) \cup \{\alpha_n\}$, and thus $\sigma(f_n) = \{\alpha_k\}_1^n$. Relative to the decomposition $H = s_n(H) \oplus (1-s_n)(H)$ we have

$$x + y_n = \begin{bmatrix} f_n & * \\ 0 & h_n \end{bmatrix}, \quad \text{where } h_n = x|(1 - s_n)(H).$$

By (BA.4.5),

$$\sigma(x + y_n) = \sigma(f_n) \cup \sigma(h_n) = (\sigma(x)\setminus\{\lambda_k\}_1^n) \cup \{\alpha_k\}_1^n = \sigma(x)\setminus\{\lambda_k\}_1^n.$$

Now if $\lambda \in \sigma(x)\setminus\omega$, then $\lambda \in \sigma(x + y_n)$ for each n, hence, since $Inv(A)$ is open in A, $\lambda \in \sigma(x + y)$. It follows that $\sigma(x)\setminus\omega \subset \sigma(x + y)$.

To prove the reverse inclusion, suppose that $\lambda \notin \sigma(x)\setminus\omega$, then $\lambda \in \rho(x) \cup \omega$, so we can choose $m \geq 1$ such that $\lambda \notin \sigma(x)\setminus\{\lambda_k\}_1^m$. Then, for $n \geq m$,

$$h_n = h_m|(1 - s_n)(H), \quad \text{and so} \quad (\lambda - h_n)^{-1} = (\lambda - h_m)^{-1}|(1 - s_n)(H),$$

where the inverses exist by virtue of the choice of m. Then, relative to the decomposition $H = s_n(H) \oplus (1 - s_n)(H)$, write

$$w_n = \begin{bmatrix} 0 & 0 \\ 0 & (\lambda - h_n)^{-1} \end{bmatrix}.$$

Then $||w_n|| \leq ||w_m||$ for $n \geq m$. Fix $n \geq m$ so that $||y - y_n|| \leq ||w_m||^{-1}$. Now, since $(y - y_n)|s_n(H) = 0$,

$$y - y_n = \begin{bmatrix} 0 & * \\ 0 & z \end{bmatrix}, \quad \text{hence} \quad w_n(y - y_n) = \begin{bmatrix} 0 & 0 \\ 0 & (\lambda - h_n)^{-1}z \end{bmatrix}.$$

$$\text{Also} \quad (\lambda - x - y_n)^{-1}(y - y_n) = \begin{bmatrix} (\lambda - f_n)^{-1} & * \\ 0 & (\lambda - h_n)^{-1} \end{bmatrix} \begin{bmatrix} 0 & * \\ 0 & z \end{bmatrix} = \begin{bmatrix} 0 & * \\ 0 & (\lambda - h_n)^{-1}z \end{bmatrix}.$$

Now, by (BA.4.5),

$$\begin{aligned} r\{(\lambda - x - y_n)^{-1}(y - y_n)\} &= r\{w_n(y - y_n)\}, \\ &\leq ||w_n|| \, ||y - y_n||, \\ &\leq ||w_m|| \, ||y - y_n|| < 1. \end{aligned}$$

Therefore $1 - (\lambda - x - y_n)^{-1}(y - y_n) \in \mathrm{Inv}(A)$. Multiplying on the left by $\lambda - x - y_n \in \mathrm{Inv}(A)$ shows that $\lambda - x - y \in \mathrm{Inv}(A)$, that is $\lambda \notin \sigma(x + y)$, therefore $\sigma(x + y) \subset \sigma(x) \setminus \omega$ ●

Now we apply this result to Riesz and Fredholm theory on A relative to a fixed closed ideal $K \subset I(A)$. But first, we identify $I(A)$.

C*.2.4 LEMMA. *In a C*-algebra* A, $I(A) = \overline{\mathrm{soc}(A)}$.

<u>Proof</u>. $A/\overline{\mathrm{soc}(A)}$ is a C*-algebra which is semisimple so, by BA.2.3,

$$\overline{\mathrm{soc}(A)} = k(h(\overline{\mathrm{soc}(A)})) = k(h(\mathrm{soc}(A))) = I(A) \quad ●$$

C*.2.5 COROLLARY. (West decomposition) *Let* x *be a Riesz element of a C*-algebra* A, *then there exists a normal* $y \in K$ *such that* $\sigma(x + y) = \{0\}$.

<u>Proof</u>. Apply C*.2.3. with $\omega = \sigma(x)\setminus\{0\}$ ●

C*.2.6 COROLLARY. (Stampfli decomposition) *Let* x *be an element of a C*-algebra* A, *then there exists* $y \in K$ *such that* $\sigma(x + y) = W(x)$ *the Weyl spectrum of* x.

<u>Proof</u>. $W(x) = \{\lambda \in \sigma(x) : \lambda - x \notin \Phi^{0}(A)\}$, so we must remove all the index-zero Fredholm points of $\sigma(x)$ by the addition of a single $y \in K$. This is done in two stages, first the 'blobs' of index-zero Fredholm points are removed one by one, then when this has been completed, C*.2.3 is applied to remove the isolated index-zero Fredholm points on $\sigma(x)$ (which have associated spectral idempotents in K (R.2.4)).

A 'blob' is a connected component of index-zero Fredholm points of $\sigma(x)$ which is not a one-point set. The blobs are countable, say, $\{\nu_n\}_1^\infty$. Take $\varepsilon_0 = 1$ and inductively construct sequences $\{\lambda_n\}_1^\infty \subset \mathbb{C}$, $\{u_n\}_1^\infty \subset K$, and $\{\varepsilon_n\}_1^\infty \subset \mathbb{R}^+$ as follows. Choose $\lambda_n \in \nu_n$, then $u_n \in K$ such that $||u_n|| \leq \frac{1}{2}\varepsilon_{n-1}$, so that

$$x + \sum_1^n u_k - \lambda_n \in \mathrm{Inv}(A),$$

(This is possible by F.3.11). Finally, choose $\varepsilon_n \leq \frac{1}{2}\varepsilon_{n-1}$ so that

$$x + \sum_1^n u_k - \lambda_n + y \in \mathrm{Inv}(A) \qquad (\dagger)$$

for $||y|| \leq \varepsilon_n$. Then

$$\Big|\Big| \sum_{n+1}^\infty u_k \Big|\Big| \leq \sum_{n+1}^\infty ||u_k|| \leq \varepsilon_n \sum_1^\infty 2^{-k} = \varepsilon_n,$$

thus $\sum_1^\infty u_k$ converges to $u \in K$. Now, by (†) we get, for each n,

$$x + u - \lambda_n \in \mathrm{Inv}(A) \iff \lambda_n \in \rho(x + u).$$

Since $\lambda_n \in \nu_n$, it follows from R.2.7, that $\sigma(x + u) \cap \nu_n$ is an at most countable set of Riesz points of $x + u$. Thus we have removed the countable set of blobs ν_n of index-zero Fredholm points of $\sigma(x)$, replacing each one by an at most countable set of Riesz points of x. We are left with the task of removing a countable set of Riesz points of $\sigma(x + u)$. So, by C*.2.3, there exists $v \in K$ such that $\sigma(x + u + v) = W(x)$ ●

C*.3 Riesz algebras

Pelczyński conjectured that if the spectrum of every hermitean element in a C*-algebra is countable, then the spectrum of every element in the algebra is countable. This conjecture has been confirmed by Huruya (47). An obvious modification leads to a characterisation of Riesz algebras among C*-algebras, which is originally due to Wong ((96) Theorem 3.1). A will denote a C*-algebra and $H(A)$ the set of *hermitean* or self-adjoint elements of A.

C*.3.1 THEOREM. *If* $\sigma(h)$ *has no non-zero accumulation point for each* $h \in H(A)$ *then* A *is a Riesz algebra.*

Proof. By virtue of C*.2.4 it is sufficient to prove that $A = \overline{soc(A)}$.

If $x \in A$, $\sigma(x^*x)$ has no non-zero accumulation point. For $\varepsilon > 0$ let p be the spectral idempotent of x^*x corresponding to the spectral set $\{\lambda \in \sigma(x^*x) : |\lambda| \geq \varepsilon^2\}$. Then $p \in H(A)$, and p commutes with x^*x, hence $||x - xp||^2 = ||(x - xp)^*(x - xp)|| = ||x^*x - px^*x|| = r(x^*x - px^*x) < \varepsilon^2$. So $||x - xp|| < \varepsilon$, and it suffices to show that $p \in soc(A)$.

Suppose that $p \notin soc(A)$ and put $p_1 = p$. Then, as in the proof of R.2.6, we construct a strictly decreasing sequence of idempotents $\{p_n\}_1^\infty$ such that, for each n, $p_n \notin soc(A)$, and, by BA.4.3, each of these idempotents may be chosen self-adjoint. Since $p \notin soc(A)$, $p \notin Min(A)$ so there exists $y \in pAp$ (which is a C*-algebra with unit p), such that $\sigma(y)$ consists of at least two points. But, a priori, y need not be in $H(A)$. If either $\sigma(y^*y)$ or $\sigma(yy^*)$ contain two points then using the hypothesis we can construct p_2 strictly less than p_1 and $p_2 \notin soc(A)$ as in R.2.6. So suppose that for each $y \in pAp$, $\sigma(y^*y)$ and $\sigma(yy^*)$ are singleton sets. If $y \neq 0$,

$$r(y^*y) = r(yy^*) = ||y^*y|| = ||y||^2 \neq 0,$$

so $\sigma_{pAp}(y^*y)$ and $\sigma_{pAp}(yy^*)$ are singleton sets, neither of which consist of the zero point. It follows that y^*y and $yy^* \in \mathrm{Inv}(pAp)$, hence $y \in \mathrm{Inv}(pAp)$. Thus pAp is a division algebra, therefore $pAp = \mathbb{C}p$, and $p \in \mathrm{Min}(A)$ which is a contradiction. Thus starting with $p_1 = p$ we can always construct an idempotent p_2 satisfying our requirements and hence, by induction, an infinite strictly decreasing sequence $\{p_n\}_1^\infty$ such that, for each n, $p_n \notin \mathrm{soc}(A)$.

Now the sequence $\{p_n\}_1^\infty$ lies in $H(A)$. Put $u_k = p_k - p_{k-1}$, then $\{u_k\}_1^\infty$ is an infinite orthogonal family of idempotents in $H(A)$. Now $u = \sum_1^\infty 2^{-k} u_k \in H(A)$, hence $p + u \in H(A)$, and 1 is an accumulation point of $\sigma(p + u)$ which contradicts the hypothesis. Therefore $p \notin \mathrm{soc}(A)$ as required ●

C*.4 A representation

We have defined finite rank and compact elements of a C*-algebra (C*.1.1). Riesz and Fredholm elements are considered relative to the closure of the socle. In this section we construct a faithful *-representation of the C*-algebra onto a closed subalgebra of the operators on a Hilbert space which maps the finite rank (respectively, compact, Riesz, Fredholm) elements onto the finite rank (respectively, compact, Riesz, Fredholm) operators in the subalgebra.

Recall that an element of an algebra is *algebraic* if it satisfies a non-trivial polynomial identity. Clearly finite rank operators on a linear space or finite rank elements in a C*-algebra are algebraic.

C*.4.1 THEOREM. *If* A *is a C*-algebra,* soc(A) *is the largest ideal of algebraic elements of* A.

Proof. $x \in \mathrm{soc}(A)$,

$\iff \dim(xAx) < \infty$, (C*.1.2)

$\Rightarrow$ x is algebraic.

Conversely, let J be an ideal of algebraic elements of A. By R.2.6, $J \subset I(A) = \overline{\mathrm{soc}(A)}$. Suppose that $x \in J \backslash \mathrm{soc}(A)$, then, by BA.4.4, $x^*x \in J \backslash \mathrm{soc}(A)$. But $x^*x = \sum_1^n \lambda_i p_i$ where $\lambda_i \in \mathbb{R}$ and $p_i = p_i^2 = p_i^*$ $(1 \le i \le n)$. Clearly some p_i (say p) $\in J \backslash \mathrm{soc}(A)$. But $p \in \overline{\mathrm{soc}(A)}$, so p

is a compact element of A (C*.1.3), that is $p_\wedge p$ is a compact operator on A which is idempotent, so $p_\wedge p$ is a finite rank operator on A, hence $p \in soc(A)$ (C*.1.2), which is a contradiction ●

The construction of our representation is done in stages. First we produce a natural family of Hilbert spaces associated with the minimal ideals of a C*-algebra.

Let A be a C*-algebra with $e = e^* \in Min(A)$, and let $H_e = Ae$ be the corresponding minimal left ideal of A. If $x, y \in H_e$ define the scalar $\langle x,y\rangle$ by

$$\langle x, y\rangle e = ey^*xe = y^*x.$$

Clearly $\langle\ ,\ \rangle$ is linear in the first variable and conjugate linear in the second. Now if $x \in H_e$,

$$(\langle x, x\rangle - ex^*xe)e = 0,$$

thus $\langle x, x\rangle \in \sigma(ex^*xe)$, and $ex^*xe = (xe)^*xe$, so $\langle x, x\rangle \geq 0$. Further

$$\langle x, x\rangle = ||\langle x, x\rangle e|| = ||ex^*xe|| = ||xe||^2 = ||x||^2,$$

thus the algebra norm on H_e is identical with the inner-product norm. It is clear that H_e is closed in A, for if $x_n \in H_e$ and $x_n \to x \in A$, then

$$x_n = x_n e \to xe = x,$$

so $x \in H_e$. Thus H_e is a Hilbert space under this inner-product. We now define a representation π_e of A on H_e as follows,

$$\pi_e(a)x = ax \qquad (a \in A,\ x \in H_e).$$

C*.4.2 LEMMA. *The representation* (π_e, H_e) *is a *-representation of* A *on* H_e *with the following properties:*

(i) π_e (span AeA) = $F(H_e)$;

(ii) $\pi_e(A) \supset K(H_e)$;

(iii) $\ker \pi_e = P_e$ *the unique primitive ideal of* A *which does not contain* e (BA.3.5).

Proof. It follows at once from the definition that π_e is a *-representation of A on H_e. If $x, y \in H_e$, let $x \otimes y$ denote the rank-one operator on H_e: $z \to \langle z, x\rangle y$ $(z \in H_e)$. Then

$$\pi_e(yx^*)z = yx^*z = yex^*ze = \langle z, x\rangle y = (x \otimes y)z,$$

thus $\pi_e(yx^*) = x \otimes y$. Now every element of AeA is of the form yx^* where $x, y \in H_e$, hence (i) follows. From this we conclude that π_e is irreducible on H_e thus $\ker(\pi_e)$ is a primitive ideal of A and since $e \notin \ker(\pi_e)$, $\ker(\pi_e) = P_e$.

(ii) follows from (i) because, since π_e is continuous, $\pi_e(A)$ is closed in $B(H_e)$ (BA.4.1) ●

In our main theorem $R(A)$ and $\Phi(A)$ will denote the set of Riesz and Fredholm elements of a C*-algebra A relative to the closure of the socle.

C*.4.3 THEOREM. *Let* A *be a C*-algebra. There exists a faithful *-(and therefore isometric) representation* (π, H) *of* A *with the following properties:*

(i) $\pi(\mathrm{soc}(A)) = F(H) \cap \pi(A)$;

(ii) $\pi(\overline{\mathrm{soc}(A)}) = K(H) \cap \pi(A)$;

(iii) $\pi(R(A)) = R(H) \cap \pi(A)$;

(iv) $\pi(\Phi(A)) = \Phi(H) \cap \pi(A)$ *if* A *is unital*.

Proof. Let Λ be a set which indexes the primitive ideals of A which do not contain soc(A). For each $\lambda \in \Lambda$, we can choose $e_\lambda = e_\lambda^* \in \mathrm{Min}(A)$ such that $P_\lambda = P_{e_\lambda}$, and then, by C*.4.2, there exists a *-representation (π_λ, H_λ) of A on H_λ. Define

$$\pi_1 = \bigoplus_{\lambda\in\Lambda} \pi_\lambda \quad \text{on} \quad H_1 = \bigoplus_{\lambda\in\Lambda} H_\lambda.$$

Then π_1 is a *-representation of A on the Hilbert space H_1. Now $\ker \pi_1 = \bigcap_{\lambda\in\Lambda} \ker(\pi_\lambda) = \bigcap_{\lambda\in\Lambda}\{P_\lambda \in \Pi(A) : P_\lambda \not\supset \mathrm{soc}(A)\}$, by C*.4.2. As π_1 may have a non-zero kernel it is necessary to add another representation π_2 in order to ensure that the sum π be faithful. Use the Gelfand-Naimark theorem ((14) 38.10) on the C*-algebra $A/\overline{\mathrm{soc}(A)}$ to construct a *-representation (π_2, H_2) of A such that $\ker(\pi_2) = \overline{\mathrm{soc}(A)}$. Put $\pi = \pi_1 \oplus \pi_2$ on $H_1 \oplus H_2$, then $\ker(\pi) = \ker \pi_1 \cap \ker \pi_2 = (0)$, so π is a faithful *-representation.

Let us examine the range of π. $e_\lambda \in P_\mu$ for $\lambda \neq \mu$, hence if $x \in \mathrm{span}(Ae_\lambda A)$, then $\pi_\mu(x) = 0$, and since $\pi_\lambda(x) \in F(H_\lambda)$, it follows that $\pi_1(x) \in F(H_1)$. Now if $x \in \mathrm{soc}(A)$, there exists a finite subset $\{\lambda_1,,,\lambda_n\}$ of Λ such that $x \in \mathrm{span}\{Ae_jA : 1 \leq j \leq n\}$, so $\pi_1(x) \in F(H_1)$. But $\ker(\pi_2) = \overline{\mathrm{soc}(A)}$, therefore $\pi(\mathrm{soc}(A)) \subset F(H)$. To verify (i), observe that the inverse image of $F(H) \cap \pi(A)$ is an ideal of algebraic elements of A, which is therefore contained in $\mathrm{soc}(A)$ (C*.4.1), therefore $\pi(\mathrm{soc}(A)) \supset F(H) \cap \pi(A)$, whence we have equality (i).

π is a faithful *-representation hence it is isometric ((75) 4.8.6), and $\pi(\overline{\mathrm{soc}(A)})$ is closed in $B(H)$, so $\overline{F(H) \cap \pi(A)} \subset \pi(\overline{\mathrm{soc}(A)}) \subset K(H) \cap \pi(A)$. To obtain equality let $T = T^* \in K(H) \cap \pi(A)$, then $T = \sum_1^\infty \lambda_i P_i$ where $\lambda_i \in \mathbb{R}$, and $P_i = P^*_i = P^2_i \in K(H) \cap \pi(A)$ for each i. But each compact projection is of finite rank, so $P_i \in F(H) \cap \pi(A)$, thus $T \in \overline{F(H) \cap \pi(A)}$. Since every operator $S \in K(H) \cap \pi(A)$ may be written $S = T_1 + iT_2$ where T_1, T_2 are self-adjoint members of $K(H) \cap \pi(A)$, it follows that $\pi(\overline{\mathrm{soc}(A)}) \supset K(H) \cap \pi(A)$, whence we have equality (ii).

The proofs of (iii) and (iv) are now straightforward (see A.1.3) ●

C*.5 Notes

Very neat proofs of the range inclusion theorems of §0.4 can be given in a Hilbert space H via the following factorisation Lemma due to Douglas (28). (The footnote in (28) announcing an extension to Banach spaces is incorrect).

C*.5.1 LEMMA. $S, T \in B(H)$, $S(H) \subset T(H) \Rightarrow$ *there exists* $C \in B(H)$ *such that* $S = TC$.

Proof. Since $S(H) \subset T(H)$, then for each $x \in H$ there exists a unique $y \in \ker(T)^\perp$ such that $Sx = Ty$. Put $Cx = y$. C is linear and we prove C

is continuous by means of the closed graph theorem ((30) p.57). Let $\{x_n\}$ be a sequence in H such that

$$\lim_n x_n = u, \quad \lim_n Cx_n = v.$$

Then there exists $y_n \in \ker(T)^\perp$ such that $Sx_n = Ty_n$ for each n, and, since $\ker(T)^\perp$ is a closed subspace of H, $\lim_n y_n = v \in \ker(T)^\perp$. So $Su = Tv$, hence $Cu = v$, thus the graph of C is closed ●

C*.5.2 LEMMA. $S, T \in \mathcal{B}(H)$, $ST = TS$ *and* $S = TC \Rightarrow r(S) \leq r(T)r(C)$.

Proof. By induction $S^n = T^nC^n$ for each n, thus $||S^n|| \leq ||T^n|| \, ||C^n||$, and the result follows from the spectral radius formula ●

C*.5.3 COROLLARY. $S \in \mathcal{B}(H)$, $T \in \mathcal{K}(H)$ *and* $S(H) \subset T(H) \Rightarrow S \in \mathcal{K}(H)$.

Proof. Apply C*.5.1 ●

C*.5.4 COROLLARY. $S \in \mathcal{B}(H)$, $T \in \mathcal{Q}(H)$, $ST = TS$ *and* $S(H) \subset T(H) \Rightarrow S \in \mathcal{Q}(H)$.

Proof. Apply C*.5.1 and C*.5.2 ●

C*.5.5 COROLLARY. $S \in \mathcal{B}(H)$, $T \in \mathcal{R}(H)$, $ST - TS \in \mathcal{K}(H)$ *and* $S(H) \subset T(H)$ $\Rightarrow S \in \mathcal{R}(H)$.

Proof. $S = TC$ by C*.5.1. Let ψ be the canonical homomorphism of $\mathcal{B}(H)$ into the Calkin algebra. Then $\psi(S), \psi(T)$ commute and $\psi(S) = \psi(T)\psi(C)$. Now $r(\psi(T)) = 0$ hence, by C*.5.2, $r(\psi(S)) = 0$, that is $S \in \mathcal{R}(H)$ ●

Alexander (4) showed that C*.1.2 is valid in a semisimple Banach algebra. Erdos (31) defined an element x of an algebra A to be *single* if $axb = 0 \Rightarrow$ either $ax = 0$ or $xb = 0$. The single elements of $\mathcal{B}(X)$ are easily seen to be the rank one operators. Making use of this concept Erdos constructs a representation of a C*-algebra similar to that in §4, see also Ylinen (100). Erdos points out that his work does not extend even to semisimple Banach algebras. In fact, in (32), Erdos, Giotopoulos and Lambrou prove that an element x of a semisimple Banach algebra has an image as a rank one operator in some faithful representation of the algebra $\Leftrightarrow x$ is single and the operator $x_\wedge x$ is compact. The representation in §4 may be used to transfer information on finite-rank, compact or Riesz operators on Hilbert space to finite-rank, compact or Riesz elements of C*-algebras. It could, for example, be used to deduce the West and Stampfli decompositions in C*-algebras (C*.2.5, C*.2.6) from their counterpart theorems for operators (C*.2.1, C*.2.2). Legg (58) has given the C*-algebra counter part of the Chui, Smith and Ward result (26) that the commutator in the West decomposition is quasinilpotent. In fact, the more detailed information on the West decomposition provided by Murphy and West (61), (see below), is all valid in a C*-algebra. Akemann and Wright (3) have further results on the wedge operator, and on the left and right regular representations in a C*-algebra. For example, they show that if $S, T \in \mathcal{B}(H)$ then $S \wedge T$ is a weakly compact operator $\Longleftrightarrow$ either S or $T \in \mathcal{K}(H)$. See also the remarks in §F.4.

Gillespie ((35), (25) p.58) constructed a Riesz operator R on a Hilbert space such that for no decomposition of R into the sum of a compact plus a quasinilpotent did these two operators commute. Chui, Smith and Ward (26) showed that if $R = K + Q$ is a West decomposition of a Riesz operator R then the commutator $[K, Q]$ is quasinilpotent. Murphy and West (61) gave a complete structure theory for the closed subalgebra (called the *decomposition* algebra) generated by K and Q. It emerges that the set of quasinilpotents forms an ideal which is equal to the radical, and that the algebra is the spatial direct sum of the radical plus the closed subalgebra generated by K.

The problem of decomposing Riesz operators on Banach spaces has been open for some time. It may even characterise Hilbert spaces up to isomorphism. Some recent progress is due to Radjavi and Laurie (73) who showed that if R is a Riesz operator on a Banach space and $\sigma(R) = \{\lambda_n\}_1^\infty$, where the eigenvalues are repeated according to algebraic multiplicity, then R has a West decomposition if $\sum_1^\infty n|\lambda_n| < \infty$.

Olsen (67) showed that if $T \in \mathcal{B}(H)$ and $T^n \in \mathcal{K}(H)$, then $T = K + Q$ where $K \in \mathcal{K}(H)$ and $Q^n = 0$. This result has been extended to C*-algebras by Akemann and Pedersen (2).

An intriguing property of the ideal of compact operators on a Hilbert space, originally due to Salinas (77), is the following.
Let $T \in \mathcal{B}(H)$ then

$$r(T + \mathcal{K}(H)) = \inf_{K \in \mathcal{K}(H)} r(T + K).$$

In fact Salinas' proof is valid in Banach spaces. This property was algebraicised by Smyth and West (87), who showed that for a large class of commutative Banach algebras, including the C*-algebras, the above property holds for every element and for every closed ideal. Pedersen (70) proved that this is true for all C*-algebras, and Murphy and West (60) gave an elementary proof. They also showed that the class of commutative Banach algebras in which this property holds for each element and for each closed ideal is comprised, roughly, of those algebras whose Gelfand transform algebra is dense in the sup-norm algebra of continuous functions on the Gelfand space. Further algebraic information on the spectral radius may be found in the elegant monograph of Aupetit (6).

The modified Pelczyński conjecture which characterises C*-Riesz algebras is due to Huruya (47) and Wong (96). The following result is stated in (27) 4.7.20, see also (10).

C*.5.6 THEOREM. *If* A *is a C*-algebra the following statements are equivalent:*

(i) A *is a Riesz algebra;*

(ii) $A = \overline{\mathrm{soc}(A)}$;

(iii) *if* J *is a closed left (resp. right) ideal of* A *then* $\mathrm{lan}(\mathrm{ran}(J)) = J$ (*resp.* $\mathrm{ran}(\mathrm{lan}(J)) = J$);

(iv) A *is *-isomorphic and isometric to a C*-subalgebra of* $K(H)$ *for some Hilbert space* H;

(v) *The Gelfand space of every maximal commutative C*-subalgebra of* A *is discrete;*

(vi) *left(resp. right) multiplication by* x *is a weakly compact operator on* A *for each* $x \in A$;

(vii) *every non-zero point of* $\sigma(x)$ *is isolated in* $\sigma(x)$ *for each* $x = x^* \in A$.

Such algebras are also called *dual* algebras.

Pelczyński's conjecture has been verified in Banach *-algebras by Kirchberg (105).

A Applications

In this chapter our general theory is applied to a number of specific examples, particularly to algebras of operators. As a consequence we shall often use operator notation but the index (unless otherwise specified) will be the index function associated with the particular algebra.

We recall first the definitions of the various spectra in R.2.1. Let A be a unital Banach algebra and K a fixed inessential ideal of A. The Fredholm spectrum of x in A is

$$\omega(x) = \{\lambda \in \mathbb{C} : \lambda - x \notin \Phi\};$$

the Weyl spectrum is

$$W(x) = \bigcap_{k \in K} \sigma(x + k);$$

while the Browder spectrum is

$$\beta(x) = \{\lambda \in \mathbb{C} : \lambda \text{ is not a Riesz point of } x\}.$$

Our applications can be classified under three main headings.

I Spectral mapping properties. The spectral mapping theorem holds for the Fredholm and Browder spectra (R.5.2) but not, in general, for the Weyl spectrum. It does, however, hold for the Weyl spectrum for triangular algebras of operators on sequence spaces and for certain quasidiagonal operators on Hilbert space.

II Lifting theorems. Suppose that $T \in \mathcal{B}(X)$ is invertible modulo $\mathcal{K}(X)$ and satisfies some additional algebraic or analytic condition. Can one find $S \in \mathrm{Inv}(\mathcal{B}(X))$ satisfying the same condition and such that $T - S \in \mathcal{K}(X)$?

III Compact perturbations. Suppose that $T \in \mathcal{B}(X)$ satisfies an algebraic or analytic condition, can one describe $\bigcap \sigma(T + K)$, where the intersection is taken over all $K \in \mathcal{K}(X)$ satisfying the same condition?

A.1 Fredholm and Riesz elements in subalgebras

We fix some notation which shall remain in effect throughout the chapter. A denotes a unital Banach algebra and B a closed subalgebra with $1 \in B \subset A$. K_A is a fixed closed ideal of A contained in $I(A)$, and $K_B = K_A \cap B$, hence by R.2.6, $K_B \subset I(B)$. We investigate the relationship between the Fredholm elements $\Phi(A)$ in A relative to K_A, and $\Phi(B)$ in B relative to K_B. Clearly $\Phi(B) \subset \Phi(A) \cap B$ but the converse does not hold in general. If we do have equality then $\omega_B(T) = \omega_A(T)$ for each $T \in B$. First we give a necessary condition if B is semisimple.

A.1.1 THEOREM. *Let* B *be semisimple. If* $\Phi(B) = \Phi(A) \cap B$ *then* $\sigma_B(T) = \sigma_A(T)$ $(T \in B)$.

Proof. It suffices to show that $Inv(A) \cap B \subset Inv(B)$. If $T \in Inv(A) \cap B$, then $T \in \Phi(A) \cap B = \Phi(B)$. Now the left and right annihilator ideals of T in A are zero, hence the same is true of the left and right annihilator ideals of T in B. By F.1.10, the left and right Barnes idempotents of T in B are both **zero**, hence $T \in Inv(B)$ ●

A.1.2 EXAMPLE. Take $A = \mathcal{B}(X)$, $K_A = \mathcal{K}(X)$ and choose $T \in \Phi(X)$ with $i_X(T) \neq 0$. Let B be the maximal commutative subalgebra of $\mathcal{B}(X)$ containing T; then $\sigma_B(S) = \sigma_A(S)$ $(S \in B)$ (BA.1.4), but $T \notin \Phi(B)$. For, if $T \in \Phi(B)$, $\iota_B(T) = 0$ since B is commutative, and we can write $T = V + K$ where $V \in Inv(B)$ and $K \in K_B$, by F.3.11, implying that $T \in \Phi^0(X)$ (the Fredholm operators of index zero in $\mathcal{B}(X)$) which is false. So the condition of Theorem A.1.1 is not sufficient for general B.

For C*-algebras we do get equality.

A.1.3 THEOREM. *Let* A *be a C*-algebra and* B *a *-subalgebra of* A; *then* $\Phi(B) = \Phi(A) \cap B$.

Proof. The map

$$\psi : T + K_B \to T + K_A \text{ of } B/K_B \to A/K_A,$$

is a *-isomorphism so $\psi(B/K_B)$ is a *-closed subalgebra of A/K_A (BA.4.1). Thus if $T \in \Phi(A) \cap B$, then $\psi(T + K_B) = T + K_A$ is invertible in A/K_A, and hence, in $\psi(B/K_B)$ (BA.4.2). Thus $T + K_B \in \mathrm{Inv}(B/K_B)$, hence $T \in \Phi(B)$ ●

Next we consider Riesz elements. $R(A)$, $R(B)$ will denote the sets of Riesz elements in A and B, respectively.

A.1.4 THEOREM. $R(B) = R(A) \cap B$, *and if* $T \in R(B)$, $\sigma_B(T) = \sigma_A(T)$.

Proof. If $T \in R(B)$, then since $K_B = B \cap K_A$, we have $T \in R(A)$ by definition.

Conversely, suppose $T \in R(A) \cap B$. Now $\sigma_A(T)$ is countable, hence $\sigma_B(T) = \sigma_A(T)$. Further, if $0 \neq \lambda \in \sigma_A(T)$, then $0 \neq P(\lambda, T) \in K_B \subset K_A$. It follows from R.2.5 that $T \in R(B)$ ●

A.1.5 THEOREM. *Let* T *be a Riesz operator on a Hilbert space* H *and let* B *be any closed unital *-subalgebra of operators on* H *containing* T. *Then* $T = K + Q$ *where* K *is a compact operator in* B, *and* Q *a quasi-nilpotent operator in* B.

Proof. This is a Corollary of A.1.4 and the West decomposition in the algebra B (C*.2.5) ●

We have the following information on the Browder spectrum.

A.1.6 LEMMA. $T \in B$ *and* $\sigma_B(T) = \sigma_A(T) \Rightarrow \beta_B(T) = \beta_A(T)$.

Proof. Similar to A.1.4 ●

An interesting consequence of this is

A.1.7 THEOREM. *Let* A *be unital and* K *be a closed inessential ideal of* A, *then, for each* $x \in A$,

$$\beta(x) = \bigcap \{\sigma(x + y) : y \in K \text{ and } xy = yx\}.$$

Proof. Set $B = \{y \in A : xy = yx\}$; then the right hand set is the Weyl spectrum of x in B, $W_B(x)$. Since $\sigma_A(x) = \sigma_B(x)$ (BA.1.4), it follows that $\beta_A(x) = \beta_B(x)$ (A.1.6). Thus it suffices to prove that $\beta_B(x) = W_B(x)$. If $\Phi^0(B) = \{x \in \Phi(B) : \iota(x) = 0\}$, then, from R.2.2,

$$W_B(x) = \{\lambda \in \mathbb{C} : \lambda - x \notin \Phi^0(B)\};$$

and

$$\beta_B(x) = \{\lambda \in \mathbb{C} : \lambda - x \notin R(B)\}.$$

Since Riesz points are automatically index-zero Fredholm points, and since isolated index-zero Fredholm points are Riesz points, it suffices to show that an index-zero Fredholm point of $\sigma_B(x)$ is isolated in $\sigma_B(x)$. Without loss of generality take $\lambda = 0$ and $x \in \Phi^o(B)$. By F.3.11, $x = v + f$ where $v \in Inv(B)$ and $f' \in soc(B')$; also $vf = fv$ by the definition of B. Let D be the closed unital (commutative) subalgebra of B generated by v, v^{-1} and f with Gelfand space Ω. The set $\{\hat{v}(\omega) : \omega \in \Omega\}$ is bounded away from zero, while $\hat{f}(\omega) = 0$, for all but at most a finite number of $\omega \in \Omega$. It follows, since $x = v + f$, that zero is an isolated point of $\sigma_D(x)$, and therefore of $\sigma_B(x)$ ●

A.2 Seminormal elements in C*-algebras

In this section A will be a C*-algebra. If $T \in A$ and $T^*T \geq TT^*$, T is *hyponormal*. If T^* is hyponormal, T is *co-hyponormal*. In either case T is called *seminormal*. We consider the Fredholm theory of these operators.

A.2.1 LEMMA. *Suppose that* $T \in \Phi(A)$:

(i) T *is hyponormal* $\Rightarrow \iota(T) \leq 0$;

(ii) T *is co-hyponormal* $\Rightarrow \iota(T) \geq 0$.

Proof. Since $\iota(T^*) = -\iota(T)$, (ii) follows from (i).

To prove (i), recalling the definition of the index function F.3.5, it suffices to consider the case of primitive A. Assume that $T \in \Phi(A)$ is hyponormal. Then there exist Barnes idempotents $P, Q \in K_A$ (which we may take to be self-adjoint (BA.4.3)), which satisfy $lan(T) = AP$ and $ran(T) = QA$. Now

$$T^*T \geq TT^* \Rightarrow 0 = QT^*TQ \geq QTT^*Q \geq 0 \Rightarrow QTT^*Q = 0 \Rightarrow$$

$$QT = 0 \Rightarrow Q \in lan(T) \Rightarrow QP = Q \Rightarrow rank(Q) \leq rank(P).$$

Thus $\iota(T) = nul(T) - def(T) = rank(Q) - rank(P) \leq 0$ ●

A.2.2 COROLLARY. *If* T *is normal*, $\iota(T) = 0$.

A.2.3 THEOREM. *If* $T \in \Phi(A)$ *is hyponormal (resp. co-hyponormal) there exists* $K \in K_A$ *such that for each non-zero* $\lambda \in \mathbb{C}$, $T + \lambda K$ *is left (resp. right) invertible.*

Proof. A consequence of A.2.1 and F.3.11 ●

B denotes a closed *-subalgebra of A.

A.2.4 THEOREM. *If* T *is seminormal and* $T \in B \cap \Phi^0(A)$ *then* $T \in \Phi^0(B)$.

Proof. $T \in \Phi(B)$ by A.1.3. By A.2.3 there exists $K \in K_B$ such that $T + K$ has either a left or a right inverse in B, and thus in A. But $\iota_A(T + K) = 0$, hence $T + K$ is invertible in A, and hence in B. Thus $\iota_B(T) = \iota_B(T + K) = 0$ ●

Next we consider the Weyl spectrum of a seminormal operator.

A.2.5 THEOREM. *Let* $A = \mathcal{B}(H)$, $K_A = \mathcal{K}(H)$ *and let* B *be a closed *-subalgebra of* A. *If* $T \in B$ *is seminormal:*

(i) $W_B(T) = W_A(T)$,

(ii) *there exists* $K \in K_A \cap B$ *such that* $W_A(T) = \sigma(T + K)$.

Proof. (i) Clearly $\lambda - T$ is seminormal for each $\lambda \in \mathbb{C}$. By R.2.2,

$$\begin{aligned} W_A(T) &= \{\lambda \in \mathbb{C} : \lambda - T \notin \Phi^0(A)\}, \\ &= \{\lambda \in \mathbb{C} : \lambda - T \notin \Phi^0(B)\}, \quad \text{(A.2.4)} \\ &= W_B(T). \end{aligned}$$

(ii) This follows from (i) and the Stampfli decomposition of T in the algebra B (C*.2.6) ●

A consequence of this result is that if L is a collection of closed subspaces of H and $\mathcal{B}_L$ is the *-closed subalgebra of $\mathcal{B}(H)$ consisting of operators which are reduced by the subspaces of L, then if T is a seminormal operator in $\mathcal{B}_L$, there exists $K_0 \in \mathcal{K}(H) \cap \mathcal{B}_L$ such that

$$\sigma(T + K_0) = \bigcap_{K \in \mathcal{K}(H)} \sigma(T + K) = W(T).$$

Next we see that Fredholm points of index zero of a seminormal operator are Riesz points.

A.2.6 THEOREM. *If* T *is a seminormal operator in* A, *then* $W_A(T) = \beta_A(T)$.

Proof. Without loss of generality take A to be primitive. Let $T \in \Phi^o(A)$; it suffices to show that either T is invertible or that zero is an isolated point of $\sigma_A(T)$. If $T^*T \geq TT^*$, as in the proof of A.2.1, there exist self-adjoint Barnes idempotents $P, Q \in K_A$ such that $QP = Q$. But then $\iota_A(T) = 0 \Rightarrow P = Q$, and an examination of F.1.11 shows that the underlying Hilbert space satisfies $H = \ker(T) \oplus T(H)$. Thus, either $T \in \mathrm{Inv}(A)$, or zero is a pole of T of finite rank ●

A.2.7 THEOREM. *Let* A *and* B *have the property that* $\sigma_B(S) = \sigma_A(S)$ $(S \in B)$. *If* $T \in B$ *is seminormal, then* $W_B(T) = W_A(T)$.

Proof. By A.2.6

$$\begin{aligned} W_A(T) &= \beta_A(T) = \beta_B(T), \quad (A.1.6) \\ &\supset W_B(T), \\ &= \bigcap_{K \in K_B} \sigma_B(T + K), \\ &\supset \bigcap_{K \in K_A} \sigma_A(T + K) \quad \text{by hypothesis}, \\ &= W_A(T) \end{aligned}$$ ●

A.3 Operators leaving a fixed subspace invariant

Let X be a Banach space and Y a fixed closed subspace of X. Put $A = \mathcal{B}(X)$ and let B be the closed subalgebra of A consisting of operators which leave Y invariant. We need preliminary information on rad(B), soc(B'), and I(B). Recall that if $T \in B$, $T|Y$ denotes the restriction of T to Y.

Define the restriction and quotient representations of B on Y and X/Y as follows:

$$\pi_r(T)y = Ty \quad (T \in B,\ y \in Y),$$

$$\pi_q(T)(x + Y) = Tx + Y \quad (T \in B,\ x \in X).$$

It is simple to check that $F(Y) \subset \pi_r(B)$, and that $F(X/Y) \subset \pi_q(B)$, thus both these representations are irreducible. Hence the ideals $P_r = \ker(\pi_r)$, $P_q = \ker(\pi_q)$ are primitive ideals of B.

A.3.1 THEOREM. (i) $\operatorname{rad}(B) = P_r \cap P_q = \{T \in B : T(Y) = (0)$ *and* $T(X) \subset Y\}$;
(ii) $\{P_r, P_q\} = \{P \in \Pi(B) : \operatorname{soc}(B') \not\subset P'\}$;
(iii) $\operatorname{soc}(B') = (F(X) \cap B)'$;
(iv) $F(X) \cap B \subset I(B)$.

Proof. (i) $J = \{T \in B : T(Y) = (0)$ and $T(X) \subset Y\}$ is a nilpotent ideal of B, so $J \subset \operatorname{rad}(B)$. But $P_r \cap P_q = J$, and $P_r, P_q \in \Pi(B)$, hence $\operatorname{rad}(B) \subset J$.
(ii) Assume that E is an element of B such that $E' \in \operatorname{Min}(B')$. The ideals P'_r and P'_q are distinct primitive ideals of B' and, by BA.3.5, E' is in either P'_r or P'_q. Moreover, E' cannot be in both, since $P'_r \cap P'_q = (0)$. Thus if $P \in \Pi(B)$, and $P \neq P_r$ or P_q, then $E' \in P'$, thus $\operatorname{soc}(B') \subset P'$.
(iii), (iv) straightforward ●

The proof of the next result is routine.

A.3.2 LEMMA. *If* $T \in \Phi(B)$ *then*

$$\iota(T)(P_r) = i_Y(\pi_r(T)),$$

and $$\iota(T)(P_q) = i_{X/Y}(\pi_q(T)).$$

Observe that we may have $T \in \operatorname{Inv}(\mathcal{B}(X))$ and $T(Y) \subset Y$ but $T|Y \notin \operatorname{Inv}(\mathcal{B}(Y))$. So, if B is also semisimple then $\Phi(B) \neq \Phi(A) \cap B$ (A.1.1).

A.3.3 THEOREM. *Let* $T \in B$ *then*
(i) $T \in \Phi(B) \iff T \in \Phi(X)$ *and* $T|Y \in \Phi(Y)$;
(ii) $T \in \Phi^0(B) \iff T \in \Phi^0(X)$ *and* $T|Y \in \Phi^0(Y)$.

Proof. (i) Suppose that $T \in \Phi(X)$ and $T|Y \in \Phi(Y)$. Choose $S \in \mathcal{B}(X)$ such that $TS - I$ and $ST - I = F \in F(X)$. Since $T|Y \in \Phi(Y)$, there exists a finite dimensional subspace Z_1 of Y such that $Y = Z_1 \oplus T(Y)$. Choose a projection $P_1 \in \mathcal{B}(X)$ with $P_1(X) = Z_1$ and $\ker(P_1) \supset T(Y)$. Again choose a closed subspace Z_2 of Y such that $Y = Z_2 \oplus (Y \cap F(Y))$, and then a projection $P_2 \in \mathcal{B}(X)$ such that $P_2(X) = F(Y)$ and $\ker(P_2) \supset Z_2$.

We verify that $(I - P_2)S(I - P_1) \in B$. If $y \in Y$, $y = z_1 + Ty_1$ where $z_1 \in Z_1$ and $y_1 \in Y$, therefore $S(I - P_1)y = STy_1 = y_1 + Fy_1$. Since $y_1 \in Y$, $y_1 = z_2 + w$ where $z_2 \in Z_2$ and $w \in Y \cap F(Y)$. Then

$$(I - P_2)S(I - P_1)y = (I - P_2)(z_2 + w + Fy_1) = z_2 \in Y.$$

Thus $(I - P_2)S(I - P_1) \in B$ and as $P_1, P_2 \in \mathcal{F}(X) \cap B$, $(I - P_2)S(I - P_1)$ is an inverse for T modulo $\mathcal{F}(X) \cap B$. Thus $T \in \Phi(B)$. The converse is obvious.

(ii) If $T \in \Phi^o(B)$, then $0 = \iota(T)(P_r) = i_Y(T|Y)$, so that $T|Y \in \Phi^o(Y)$. But, (0.2.8), there exists $F \in \mathcal{F}(X) \cap B$ such that $T + F \in \mathrm{Inv}(B)$. Thus $i_X(T) = i_X(T + F) = 0$, giving $T \in \Phi^o(X)$.

Conversely, let $T \in \Phi^o(X)$ and $T|Y \in \Phi^o(Y)$; then, by (i), $T \in \Phi(B)$. Further $\iota(T)(P_r) = i_Y(T|Y) = 0$, by hypothesis. Suppose $\iota(T)(P_q) \leq 0$ (the case $\iota(T)(P_q) \geq 0$ is similar). Since $\iota(T) \leq 0$, there exists $F \in \mathcal{F}(X) \cap B$ such that $T + F$ has a left inverse $S \in B$ (F.3.11). Thus $T + K$ is left invertible in $\mathcal{B}(X)$. But since $i_X(T + F) = i_X(T) = 0$, $T + F \in \mathrm{Inv}(\mathcal{B}(X))$. The inverse of $T + F$ must be S thus $T + F \in \mathrm{Inv}(B)$ hence $T \in \Phi^o(B)$ ●

The next result is a Corollary of F.3.11 and A.3.3.

A.3.4 THEOREM. *Let* $T \in \mathcal{B}(X)$ *and* $T(Y) \subset Y$. *Then,* $T = V + F$ *where* $V \in \mathrm{Inv}(\mathcal{B}(X))$, *and* Y *is invariant under* V, V^{-1} *and* $F \Longleftrightarrow T \in \Phi^o(X)$ *and* $T|Y \in \Phi^o(Y)$.

A.4 Triangular operators on sequence spaces

In this section X will denote one of the sequence spaces c_o or ℓ_p $(1 \leq p \leq \infty)$ and $\{e_n\}_1^\infty$ will be the usual Schauder basis for X. If $x \in X$, $\alpha \in X'$ put $\langle x, \alpha\rangle = \alpha(x)$ and $\alpha_n = \alpha(e_n)$. Then $\langle x, \alpha\rangle = \sum_1^\infty x_n\alpha_n$ where $x = \sum_1^\infty \alpha_n e_n$. If $T \in \mathcal{B}(X)$ the corresponding matrix $[t_{ij}]$ is defined by $t_{ij} = \langle Te_j, e_i\rangle$ $(1 \leq i, j \leq \infty)$, and for convenience we write $t_i = t_{ii}$. $T \in \mathcal{B}(X)$ is *upper-triangular* if $t_{ij} = 0$ for $i > j$.

In this section $A = \mathcal{B}(X)$, $K_A = \mathcal{K}(X)$, and B denotes the closed subalgebra of A of upper-triangular operators. It is easy to check that $\mathrm{Inv}(B) = \mathrm{Inv}(A) \cap B$. The first lemma is elementary.

A.4.1 LEMMA. *Suppose that* $T \in B$ *and that* $t_i \neq 0$ $(i \geq 1)$. *If* $\{\lambda_i\}_1^\infty$ *is a sequence in* $\mathbb{C}$ *such that* $\sum_{i=1}^\infty \lambda_i t_{ij} = 0$ $(j \geq 1)$, *then* $\lambda_i = 0$ $(i \geq 1)$.

A.4.2 LEMMA. *Suppose that* $T \in B$ *and that* $t_i \neq 0$ $(i \geq 1)$, *then* $T(X)$ *is dense in* X, *and, if* $T \in \Phi^0(X)$, T *is invertible.*

Proof. If $\alpha(T(X)) = 0$ for some $\alpha \in X'$, then

$$0 = \langle Te_n, \alpha\rangle = \sum_{i=1}^\infty \alpha_i t_{in} \quad (n \geq 1).$$

Thus $\alpha_i = 0$ $(i \geq 1)$, by A.4.1, hence $\alpha = 0$, and $T(X)$ is dense in X. If, in addition, $T \in \Phi^0(X)$ then $T(X) = \overline{T(X)} = X$ so $d(T) = 0$. But $0 = i_X(T)$, so $n(T) = 0$, hence, by F.2.8, T is invertible ●

A.4.3 LEMMA. $T \in B \cap \Phi^0(X) \Rightarrow t_i = 0$ *for at most a finite number of indices* i.

Proof. Suppose that the set $W = \{i : t_i = 0\}$ is infinite. Choose $\varepsilon > 0$ such that $S \in \mathcal{B}(X)$ and $||S|| < \varepsilon \Rightarrow T + S \in \Phi^0(X)$. Take S to be the operator corresponding to the diagonal matrix $[s_{ij}]$ where $s_{ij} = 0$ $(i \neq j)$, $s_{ii} = 0$ $(i \notin W)$ and $s_{ii} = \varepsilon i^{-1}$ $(i \in W)$. Then $S \in B$ and $||S|| < \varepsilon$, thus $T_1 = T + S \in B \cap \Phi^0(X)$, and, by construction, all the diagonal entries of the matrix T_1 are non-zero. By A.4.2, T_1 is invertible, but its diagonal entries are not bounded away from zero which gives a contradiction ●

If $T \in B$ let T_Δ denote the *diagonal* operator whose diagonal entries are those of T.

A.4.4 THEOREM. $\Phi(B) = \Phi^0(B) = B \cap \Phi^0(X)$.

Proof. Suppose that $T \in B \cap \Phi^0(X)$. By A.4.3, $W = \{i : t_i = 0\}$ is finite. Define S to be the diagonal operator with diagonal entries $\{s_i\}_1^\infty$ where $s_i = 1$ $(i \in W)$ and $s_i = 0$ $(i \notin W)$. If $T_1 = T + S$, since $S \in B \cap K(X)$, $T_1 \in B \cap \Phi^0(X)$, also the diagonal entries of T_1 are all non-zero so, by A.4.2, T is invertible in $\mathcal{B}(X)$, and hence in B. Thus $T = T_1 - S \in \Phi^0(B)$

so $B \cap \Phi^{o}(X) \subset \Phi^{o}(B)$.

Now suppose that $T \in \Phi(B)$. Then there exists $S \in B$ and $K, L \in K_B$ such that $ST = I + K$, $TS = I + L$. Hence $S_\Delta T_\Delta = I + K_\Delta$, $T_\Delta S_\Delta = I + L_\Delta$. This implies that T_Δ is a Fredholm element of the commutative Banach algebra Δ of all diagonal operators on X relative to the ideal $K_\Delta = K(X) \cap \Delta$. Hence $\iota_\Delta(T_\Delta) = 0$ (where ι_Δ denotes the index function in the algebra Δ) and so $T_\Delta = V + M$ where $V \in \mathrm{Inv}(\Delta)$, and $M \in K_\Delta$, by F.3.11. If we put $R = T - M$, then $R_\Delta = T_\Delta - M = V \in \mathrm{Inv}(B)$ since $\Delta \subset B$. So $R_\Delta \in \Phi(B)$ and all its diagonal entries are non-zero. By A.4.2, $R(X) = \overline{R(X)} = X$. Hence $\iota(T) = \iota(R + M) = \iota(R) \geq 0$. The same argument applied to S gives $\iota(S) \geq 0$. But $\iota(S) = -\iota(T)$, thus $\iota(T) = 0$. Thus we have $B \cap \Phi^{o}(X) \subset \Phi^{o}(B) = \Phi(B)$ ●

A.4.5 COROLLARY. *If* T *is an upper triangular operator on* X *and* $f \in \mathrm{Hol}(\sigma(T))$ *then* $W(f(T)) = f(W(T))$.

<u>Proof</u>. $W(T) = \omega_B(T) = \sigma(T + K_B)$ $(T \in B)$, by A.4.4. Also $T \in B \Rightarrow f(T) \in B$, and the result follows from the spectral mapping theorem for ω_B (R.5.2) ●

A.5 <u>Algebras of quasitriangular operators</u>

If H is a Hilbert space let P denote the set of hermitean projections in $F(H)$ ordered by $P \leq Q$ if $PQ = QP = P$ $(\Longleftrightarrow P(H) \subset Q(H))$ for $P, Q \in P$. $T \in B(H)$ is *quasitriangular* if

$$\liminf_{P} \|PTP - TP\| = 0,$$

and the set of quasitriangular operators in $B(H)$ is denoted by Q_Δ. These operators were first studied by Halmos (39) who showed that $K(H) \subset Q_\Delta = \overline{Q}_\Delta$. Note that Q_Δ is not an algebra. Let $A = B(H)$ and $K_A = K(H)$.

A.5.1 LEMMA. $T \in Q_\Delta \cap \Phi(H) \Rightarrow i_H(T) \geq 0$.

<u>Proof</u>. Suppose, on the contrary, that $i_H(T) < 0$. By F.3.11, there exists an $F \in F(H)$ such that $T + F$ has a left inverse S, and a $P \in P$, $0 \neq P$ such that $P(T + F) = 0$. Put $R = T + F$, then $R \in Q_\Delta$, since $Q_\Delta + K(H) \subset Q_\Delta$. Let $Q \in P$ with $Q \geq P$; then $PQRQ = 0$, and since

$Q\mathcal{B}(H)Q$ is finite dimensional (C*.1.2), there exists a $Q_o \in \mathcal{P}$, $Q_o \leq Q$ such that $QRQQ_o = 0$. (To verify this observe that $0 \neq P \in Q\mathcal{B}(H)Q$, and that $P(QRQ) = 0$. So QRQ is not left invertible in the algebra $Q\mathcal{B}(H)Q$, and, since this algebra is finite dimensional, QRQ is not right invertible therein. So there is a non-zero $Q_o \in Q\mathcal{B}(H)Q$ such that $QRQQ_o = 0$, and we can choose Q_o to be a projection which is therefore $\leq Q$). Now

$$||S||\cdot||RQ - QRQ||\cdot||Q_o|| \geq ||S(RQ - QRQ)Q_o|| = ||Q_o|| = 1.$$

Thus $||RQ - QRQ|| \geq ||S||^{-1}$ (for $Q \in \mathcal{P}$ such that $Q \geq P$), contradicting the fact that $R \in \mathcal{Q}_\Delta$ ●

Now let $A = \mathcal{B}(H)$, $K_A = \mathcal{K}(H)$ and let B be any unital inverse-closed C*-subalgebra of $\mathcal{Q}_\Delta$ ($T, T^{-1} \in \mathcal{B}(H)$ and $T \in B \Rightarrow T^{-1} \in B$) which contains $\mathcal{K}(H)$, with $K_B = K_A \cap B$. Note that in such an algebra B, the index function of $T \in B$ has the property that $\iota_B(T)(P) = 0$ for $P \in \Pi(B)$ except perhaps at the zero ideal $\{0\}$, where $\iota_B(T)(\{0\}) = i_H(T)$. First let us see that such algebras B exist.

A.5.2 EXAMPLE. Let H be separable and fix an increasing sequence $P_n \in \mathcal{P}$ such that $P_n \to I$ strongly. Define B by

$$B = \{T \in \mathcal{B}(H) : ||P_nT - TP_n|| \to 0 \quad (n \to \infty)\}.$$

Routine computation shows that B is a closed *-subalgebra of $\mathcal{B}(H)$. Let $x, y \in H$, then

$$\begin{aligned}||P_n(x \otimes y) - (x \otimes y)P_n|| &= ||x \otimes (P_ny) - (P_nx) \otimes y||,\\ &\leq ||x \otimes (P_ny) - x \otimes y|| + ||x \otimes y - (P_nx) \otimes y||,\\ &\leq ||x||\cdot||P_ny - y|| + ||x - P_nx||\cdot||y|| \to 0\\ &\qquad (n \to \infty).\end{aligned}$$

Hence $x \otimes y \in B$, and it follows that $\mathcal{K}(H) \subset B$. Since B is a unital C*-subalgebra of A, B is inverse closed (BA.4.2).

A.5.3 THEOREM. (i) $\Phi(B) = \Phi^{o}(B) = \Phi^{o}(H) \cap B$.

(ii) $T \in B \Rightarrow W_B(T) = W(T) = \omega_B(T) = \omega(T)$.

Proof. (i) If $T \in \Phi(B)$, there exists $S \in B$ such that $ST - I$, $TS - I \in K_B \subset K(H)$. Hence $S, T \in \Phi(H)$ and, by A.5.1, $i_H(T), i_H(S) \geq 0$; but $i_H(S) = -i_H(T)$ so $i_H(T) = 0$. Thus $T \in \Phi^{o}(H)$. Hence $T = V + K$, where $V \in \text{Inv}(\mathcal{B}(H))$ and $K \in K(H)$ (0.2.8). But the hypothesis on B implies that $V \in \text{Inv}(B)$, therefore $\iota_B(T) = 0$, and we have shown that $\Phi(B) \subset B \cap \Phi^{o}(H) \subset \Phi^{o}(B)$. This proves (i), and (ii) is an easy consequence ●

$T \in \mathcal{B}(H)$ is *quasidiagonal* if

$$\liminf_{P} ||TP - PT|| = 0 .$$

A.5.4 COROLLARY. *If* T *is quasidiagonal and if* $f \in \text{Hol}(\sigma(T))$ *then* $f(W(T)) = W(f(T))$.

Proof. If T is quasidiagonal it is quasitriangular, hence there exists a C*-subalgebra B containing $K(H)$ and T. Then the result follows from A.5.3 and R.5.2 ●

Note that T normal, K compact $\Rightarrow T + K$ is quasidiagonal, so this result applies to a large class of operators in $\mathcal{B}(H)$.

A.6 Measures on compact groups

The background for this section may be found in (45). Let G be a compact group and $M(G)$ the convolution algebra of complex regular Borel measures on G. $\Sigma(G)$ denotes the set of equivalence classes of irreducible strongly continuous unitary representations of G and $T(G)$ the set of all trigonometric polynomials on G.

For $\sigma \in \Sigma(G)$ let $\chi_\sigma(x) = \text{tr}(\sigma(x))$ be the corresponding character; then χ_σ is a central function in $L^1(G)$ and $M_\sigma = \chi_\sigma * L^1(G)$ is a finite dimensional minimal ideal in $M(G)$. There exists a constant $d_\sigma > 0$ such that $e_\sigma = d_\sigma \chi_\sigma$ is the identity of M_σ. As usual we identify $L^1(G)$ with the set of measures in $M(G)$ which are absolutely continuous with respect to Haar measure on G. Note that $\overline{T(G)} = L^1(G)$.

A.6.1 LEMMA. $T(G) = \text{soc}(M(G))$.

Proof. Since M_σ is finite dimensional, $M_\sigma \varepsilon$ soc(M(G)) for each σ, hence $\mathcal{T}(G) = \text{span}\{M_\sigma : \sigma \varepsilon \Sigma(G)\} \subset \text{soc}(M(G))$.

If $\mu \varepsilon M(G)$, let $\hat{\mu}(\sigma)$ $(\sigma \varepsilon \Sigma(G))$ be a fixed Fourier-Stieltjes transform of μ. Let $e \varepsilon \text{Min}(M(G))$, then there exists $\sigma \varepsilon \Sigma(G)$ such that $\hat{e}(\sigma) \neq 0$ ((45) 28.36). Hence $(e*\chi_\sigma)\hat{}\,(\sigma) \neq 0$ ((45) 28.39), thus $e*\chi_\sigma \neq 0$ and $e*\chi_\sigma \varepsilon \mathcal{T}(G)$. It follows that $\text{soc}(M(G)) \subset \mathcal{T}(G)$ ●

Thus $L^1(G) = \overline{\text{soc}(M(G))}$ is a closed ideal of $M(G)$ which is a Riesz algebra. Let $\Phi(M(G))$ denote the set of Fredholm elements in $M(G)$ relative to $L^1(G)$. If $\mu \varepsilon M(G)$ define $T_\mu \varepsilon \mathcal{B}(L^1(G))$ by $T_\mu x = \mu * x$ $(x \varepsilon L^1(G))$. Let δ_o be the identity measure on $M(G)$.

A.6.2 LEMMA. $\mu \varepsilon M(G) \Rightarrow \sigma(T_\mu) = \sigma(\mu)$.

Proof. Suppose $T_\mu \varepsilon \text{Inv}(\mathcal{B}(L^1(G)))$, then there exists $S \varepsilon \mathcal{B}(L^1(G))$ such that $T_\mu S = T_{\delta_o} = ST_\mu$. If $x, y \varepsilon L^1(G)$, then $T_\mu((Sx)*y) = \mu*(Sx)*y = (T_\mu Sx)*y = x*y = T_\mu(S(x*y))$. Thus $(Sx)*y = S(x*y)$ $(x,y \varepsilon L^1(G))$. By Wendel's Theorem ((45) 35.5), $S = T_\nu$, for some $\nu \varepsilon M(G)$, thus $\nu = \mu^{-1}$ in $M(G)$.

A.6.3 THEOREM. $\mu*L^1(G)$ *has finite co-dimension in* $L^1(G) \Leftrightarrow T_\mu$ *is a Riesz-Schauder operator.*

Proof. $\mu*L^1(G) = T_\mu(L^1(G))$ hence, by (25) 3.2.5, since $\mu*L^1(G)$ has finite co-dimension it is closed in $L^1(G)$. Suppose that $\{\sigma_1, \ldots, \sigma_m\}$ is a set of distinct elements of $\Sigma(G)$, and that there exist $y_k \varepsilon M_{\sigma_k}$, $y_k \notin \mu*L^1(G)$ $(1 \leq k \leq m)$. If $\lambda_1 y_1 + \ldots + \lambda_m y_m = 0$ where $\lambda_k \varepsilon \mathbb{C}$ $(1 \leq k \leq m)$, then

$$\lambda_j y_j = (\lambda_1 y_1 + \ldots + \lambda_m y_m)*e_{\sigma_j} = 0.$$

Thus $\{y_1, \ldots, y_m\}$ is a linearly independent set. It follows that, for $\sigma \varepsilon \Sigma(G)$, with the possible exception of a finite subset $E \subset \Sigma(G)$, $M_\sigma \subset \mu*L^1(G)$. If $\sigma \varepsilon \Sigma(G)\backslash E$ and $x \varepsilon M_\sigma$, there exists $y \varepsilon L^1(G)$ such that $\mu*y = x$. Hence $\mu*(e_\sigma*y) = e_\sigma*x = x$. Thus $\mu*M_\sigma = M_\sigma$ $(\sigma \varepsilon \Sigma(G)\backslash E)$. Put $X_o = \text{span}\{M_\sigma : \sigma \varepsilon \Sigma(G)\backslash E\}$, $X = \bar{X}_o$ and $e = \sum_{\sigma \varepsilon E} e_\sigma$ (if $E = \phi$ put $e = 0$). Then T_e is the projection of $L^1(G)$ onto $\sum_{\sigma \varepsilon E} M_\sigma$ and $\ker(T_e) = X$. Also $T_e T_\mu = T_\mu T_e$. Now since $\mu*M_\sigma = M_\sigma$ $(\sigma \varepsilon \Sigma(G)\backslash E)$, then $X_o \subset \mu*L^1(G)$, and hence $X \subset \mu*L^1(G)$. Therefore $X_o = (\delta_o - e)*\mu*L^1(G) = \mu*X$. Hence $T_\mu(X) = X$. Let $Y = \{x \varepsilon L^1(G) : \mu*x = 0\}$. Since M_σ is always finite dimensional and $\mu*M_\sigma = M_\sigma$ $(\sigma \notin E)$, it follows that $Y \cap M_\sigma = (0)$ $(\sigma \notin E)$. Then $Y*M_\sigma \subset Y \cap M_\sigma{}^2 = (0)$ if $\sigma \notin E$ so $Y \subset \text{lan}(X_o)$. Thus $Y \subset \text{lan}(X)$. Therefore $(Y \cap X)^2 = (0)$, hence $Y \cap X = (0)$, since $L^1(G)$ is semisimple.

We have proved that $\ker(T_\mu) \cap X = (0)$ and that $T_\mu(X) = X$. Thus $T_\mu | X \in \text{Inv}(\mathcal{B}(X))$, hence $S = T_e + (I - T_e)T_\mu \in \text{Inv}(\mathcal{B}(L^1(G)))$. Now $K = T_e(T_\mu - I) \in \mathcal{F}(L^1(G))$, and $T_\mu = S + K$ with $SK = KS$. Thus T_μ is a Riesz-Schauder operator ●

A.6.4 COROLLARY. *If* $\mu \in M(G)$ *the following are equivalent:*

(i) $\mu*L^1(G)$ *has finite co-dimension in* $L^1(G)$;

(ii) $\mu \in \Phi(M(G))$;

(iii) $\mu \in \Phi^o(M(G))$;

(iv) $\mu = \nu + \kappa$ *where* $\nu \in \text{Inv}(M(G))$, $\kappa \in \mathcal{T}(G)$ *and* $\mu*\kappa = \kappa*\mu$;

(v) $\mu = (\delta_o - \phi)*\nu$ *where* $\nu \in \text{Inv}(M(G))$ *and* $\phi = \phi^2 \in \mathcal{T}(G)$.

Proof. (i) ⟹ (iv) by A.6.3, and (iv) ⟹ (iii) ⟹ (ii) clearly. If $\mu \in \Phi(M(G))$, then $T_\mu \in \Phi(L^1(G))$, hence (ii) ⟹ (i).

Obviously (v) ⟹ (iii). Conversely, if $\mu \in \Phi^o(M(G))$ then, by F.3.11, there exists $\kappa \in \phi_1*M(G)*\phi_2$ such that $\mu + \kappa \in \text{Inv}(M(G))$, $\mu*\phi_1 = 0 = \phi_2*\mu$, and ϕ_1, ϕ_2 are idempotents in $\text{soc}(M(G)) = \mathcal{T}(G)$ (A.6.1). Now $\mu = (\delta_o - \phi_1)*(\mu + \kappa)$ is a factorisation of μ as in (v) ●

A.6.5 COROLLARY. $\mu \in M(G) \Rightarrow \omega(\mu) = W(\mu) = \beta(\mu)$.

Proof. A.6.4 implies that if $\lambda\delta_o - \mu \in \Phi(M(G))$, then λ is a Riesz point of μ ((25) 1.4.5) ●

A.7 Notes

I Spectral mapping properties. Gramsch and Lay (38) prove spectral mapping theorems for essential spectra in a general context. Assume that S is an open semigroup in a unital Banach algebra A which contains $\text{Inv}(A)$ and for $x \in A$ define

$$\sigma_S(x) = \{\lambda \in \mathbb{C} : \lambda - x \notin S\}.$$

Then Gramsch and Lay say that the spectral mapping theorem holds in this context if for each $f \in \text{Hol}(\sigma_s(x))$ and $x \in A$

$$\sigma_s(f(x)) = f(\sigma_s(x)).$$

They show that the spectral mapping theorem holds for a number of essential spectra of interest in operator theory including those of Bowder (16) ($\beta(x)$ in our notation), Kato (52) and Schechter (79) and an example is given to show that it fails for the Weyl spectrum $W(x)$.

II <u>Lifting theorems</u>. The decomposition theorems of West and Stampfli, proved in §C*.2, extend the corresponding lifting theorems from the Calkin algebra to a C*-algebra setting. This observation applies to the results of Olsen and Pedersen mentioned in §C*.5. For a Banach space X the extension of the lifting theorem $\Phi^{o}(X) = \mathrm{Inv}(\mathcal{B}(X)) + \mathcal{K}(X)$ due to Pearlman and Smyth has been discussed in §F.3. Lay (56) pointed out that if $T = \Phi^{o}(X)$ has a commuting decomposition into the sum of an invertible and a finite rank operator then zero is a Riesz point of T (the Riesz-Schauder case).

III <u>Compact perturbations</u>. If $T \in \mathcal{B}(X)$, Schechter (79) shows that $\{\lambda \in \mathbb{C} : \lambda - T \notin \Phi^{o}(X)\}$ is the largest subset of $\sigma(T)$ which is invariant under all compact perturbations of T; equivalently,

$$\{\lambda \in \mathbb{C} : \lambda - T \notin \Phi^{o}(X)\} = \bigcap_{K \in \mathcal{K}(X)} \sigma(T + K).$$

The generalisation of this result to Banach algebras is given in R.2.2. Lay (56) proved that

$$\beta(T) = \bigcap\{\sigma(T + K) : K \in \mathcal{K}(X) \text{ and } KT = TK\}.$$

The verification of this result in Banach algebras is contained in the proof of R.5.2.

BA Banach algebras

This chapter lists the information required from algebra theory, and in particular deals with Banach algebras over the complex field. Where the results are known they will be referenced in one of the standard texts (14), (75), (48). Otherwise proofs are given. The algebras will always be unital and complex, the non-unital case may be dealt with by adjoining an identity.

§1 deals with basic spectral theory and §2 with the space of primitive ideals in the hull-kernel topology, in the commutative case with the space of maximal ideals. §3 gathers information on minimal ideals and the socle, while basic results on C*-algebras are listed in §4.

BA.1 Spectral theory

Let A be a unital Banach algebra and let Inv(A) denote the set of invertible elements in A. If $x \in A$, the resolvent set $\rho(x)$ of x in A is the set

$$\rho(x) = \rho_A(x) = \{\lambda \in \mathbb{C} : \lambda - x \in \mathrm{Inv}(A)\},$$

while the spectrum $\sigma(x) = \sigma_A(x) = \mathbb{C}\setminus\rho(x)$. The subscripts may be omitted if the algebra in question is unambiguous.

BA.1.1 Inv(A) *is an open subset of* A *and if* $x \in A$, $\sigma_A(x)$ *is a non-empty compact subset of* $\mathbb{C}$.
((14) 2.12 and 5.8).

We use ∂ to denote the topological boundary of a set.

BA.1.2 *Let* B *be a closed subalgebra of* A *containing* 1, x. *Then* $\sigma_B(x) \supset \sigma_A(x)$, *while* $\partial\sigma_B(x) \subset \partial\sigma_A(x)$.
((14) 5.12).

For $x \in A$ the spectral radius is defined by

$r(x) = \sup\{|\lambda| : \lambda \in \sigma(x)\}.$

BA.1.3 $r(x) = \lim_{n\to\infty} ||x^n||^{1/n}.$

((14) 2.8). It follows that the spectral radius is independent of the containing algebra.

BA.1.4 *If* B *is a maximal commutative subalgebra of* A *containing* x *it will be unital and* $\sigma_B(x) = \sigma_A(x)$.
((14) 15.4).

If σ is a non-empty compact subset of $\mathbb{C}$, denote by Hol(σ) the class of complex valued functions which are defined and holomorphic on some neighbourhood of σ. Hol(σ) may be regarded as an algebra if we restrict a combination of functions to the intersection of their domains. If $x \in A$ and $f \in \text{Hol}(\sigma(x))$, then an element $f(x) \in A$ is defined by means of an A-valued Cauchy integral ((14) §7).

BA.1.5 *If* $x \in A$ *the map* $f \to f(x)$ *is an algebra homomorphism of* Hol($\sigma(x)$) *into* A, *mapping complex polynomials into the corresponding polynomials in* x.
((14) 7.4).

BA.1.6 (Spectral mapping theorem). *If* $x \in A$ *and* $f \in \text{Hol}(\sigma(x))$, *then* $\sigma(f(x)) = f(\sigma(x))$.
((14) 7.4).

Note that if $x \in A$ satisfies $||1 - x|| < 1$, then we can define $\log x$ by means of an A-valued Cauchy integral.

BA.2 The structure space

From BA.2.1 to BA.2.5 we shall be dealing with a purely algebraic situation. Banach algebras are introduced after 2.5. The term ideal without either adjective right, or left, means a two-sided ideal.

Let A be a unital algebra and X a linear space, a *representation* of A on X is an algebra homomorphism of A into the algebra $L(X)$ of linear operators on X. A representation $\phi : A \to L(X)$ is (strictly or algebraically) *irreducible* if $\phi \neq 0$, and if the only subspaces which are invariant under $\phi(x)$ for each $x \in A$ are (0) and X. An ideal P of A is *primitive* if there exists a maximal modular left ideal L of A such that $P = \{x \in A : xA \subset L\}$.

BA.2.1 (i) P *is a primitive ideal of* A $\Leftrightarrow$ P *is the kernel of an irreducible representation of* A;

(ii) P *is the kernel of the (irreducible) left regular representation on the quotient space* A/L;

(iii) *If* P *is a primitive ideal of* A *and* L_1, L_2 *are left ideals of* A *with* $L_1L_2 \subset P$ *then either* $L_1 \subset P$ *or* $L_2 \subset P$.

((14) 24.12). Note that L_1L_2 is made up of elements $\sum_1^n x_iy_i$ $(x_i \in L_1, y_i \in L_2)$.

The set of all primitive ideals of A is denoted by $\Pi(A)$ and has a standard topology known as the *hull-kernel topology* defined in terms of a closure operation. If $\Gamma \subset \Pi(A)$ the *kernel* of Γ is $k(\Gamma) = \bigcap\{P \in \Pi(A) : P \in \Gamma\}$ while if V is a non-empty subset of A the *hull* of V is $h(V) = \{P \in \Pi(A) : P \supset V\}$. A set $\Gamma \subset \Pi(A)$ is defined to be closed if $\Gamma = h(k(\Gamma))$ ((14) p.132). The *radical* of A, $\text{rad}(A) = \bigcap\{P : P \in \Pi(A)\}$. If $\Pi(A)$ is empty, rad(A) = A. A is *semisimple* if rad(A) = (0). We collect some information on the radical. An ideal J such that $J^n = (0)$ is *nilpotent*.

BA.2.2 (i) A/rad(A) *is semisimple;*

(ii) *if* J *is an ideal of* A, $\text{rad}(J) = J \cap \text{rad}(A)$;

(iii) *if* J *is a nilpotent left or right ideal of* A *then* $J \subset \text{rad}(A)$;

(iv) $z \in \text{rad}(A) \Rightarrow 1 + z \in \text{Inv}(A)$;

(v) $x \in \text{Inv}(A) \Leftrightarrow x + P \in \text{Inv}(A/P)$ $(P \in \Pi(A))$.

Proof. For (i) - (iv) see (14) 24.16, 24.20, 24.21.

(v) $\Rightarrow$ is obvious. So suppose x + P is invertible for $P \in \Pi(A)$. We show that Ax = A. If not then Ax is a left ideal which is contained in a maximal left ideal L. Then $Q = \{a \in A : aA \subset L\} \in \Pi(A)$, and, by hypothesis, there exists $u \in A$ such that $ux - 1 \in Q \subset L$. But $ux \in Ax \subset L$, hence $1 \in L$ which is a contradiction. So Ax = A and x has a left inverse $y \in A$. Thus x + P has y + P as a left inverse, and therefore as its inverse in A/P $(P \in \Pi(A))$. Now the hypothesis holds for y so, as above, y has a left inverse in A. Therefore $y = x^{-1}$ ●

Let J be an ideal of A and let $\phi : A \to A/J$ be the canonical homomorphism.

BA.2.3 (i) $P \in h(J) \Longleftrightarrow \phi(P) \in \Pi(A/J)$;
(ii) $k(h(J)) = \phi^{-1}(\mathrm{rad}(A/J))$;
(iii) A/J *is semisimple* $\Longleftrightarrow J = k(h(J))$.
((48) p.205).

The next result is frequently used in the text.

BA.2.4 *Let* A *be a unital algebra with ideal* J, *then there exists* $y \in A$ *such that* $xy - 1, yx - 1 \in J \Longleftrightarrow$ *there exists* $y \in A$ *such that* $xy - 1$, $yx - 1 \in k(h(J))$.

Proof. $\Rightarrow$ is obvious, so let ϕ denote the canonical homomorphism $\phi : A \to A/J$.

x is invertible modulo $k(h(J))$

$\Rightarrow$ there exists $y \in A$ such that $xy - 1, yx - 1 \in k(h(J))$,

$\Rightarrow \phi(x)\phi(y) - \phi(1), \phi(y)\phi(x) - \phi(1) \in \phi(k(h(J))) = \mathrm{rad}(A/J)$ (BA.2.3),

$\Rightarrow \phi(x)\phi(y), \phi(y)\phi(x) \in \mathrm{Inv}(A/J)$ (BA.2.2),

$\Rightarrow \phi(x) \in \mathrm{Inv}(A/J)$ ●

If $x \in A$ set $x' = x + \mathrm{rad}(A) \in A' = A/\mathrm{rad}(A)$.
If $W \subset A$ set $W' = \{x' : x \in W\}$.

BA.2.5 (i) *The structure spaces* $\Pi(A)$, $\Pi(A')$ *are homeomorphic under the map* $P \to P'$;
(ii) x *has a left (right) inverse in* $A \Longleftrightarrow x'$ *has a left(right) inverse in* A'.
(iii) $\mathrm{Inv}(A)' = \mathrm{Inv}(A')$.

Proof. (i) ((14) 26.6).
(ii) $\Rightarrow$ is obvious, so assume that x' has a left inverse in A'. Then there exist $y \in A$, $z \in \mathrm{rad}(A)$ such that $yx = 1 + z$. But $1 + z \in \mathrm{Inv}(A)$ (BA.2.2), so x has a left inverse in A.
(iii) follows at once ●

Now specialise to the case of a Banach algebra A. Let $P \in \Pi(A)$ then there exists a maximal modular (and therefore closed) left ideal L of A such that $P = \{x \in A : xA \subset L\}$. It follows that P is closed in A.

Further, by BA.2.1, P is the kernel of the irreducible left regular representation on the quotient space A/L which is a Banach space. Now the image of this representation is contained in $\mathcal{B}(A/L)$, the bounded linear operators on A/L, hence, by Johnson's theorem ((14) 25.7), this representation is continuous. Thus, without loss of generality, when dealing with Banach algebras it will be sufficient to consider the continuous irreducible representations into $\mathcal{B}(X)$ for Banach spaces X.

If A is a Banach algebra, then $rad(A)$ is a closed ideal of A, and $A' = A/rad(A)$ is a semisimple Banach algebra. It follows from BA.2.5 that $\sigma_A(x) = \sigma_{A'}(x')$ $(x \in A)$. An algebra A is *primitive* if zero is a primitive ideal of A.

BA.2.6 *If* A *is a Banach algebra and* $P \in \Pi(A)$ *the primitive Banach algebras* A/P *and* A'/P' *are isometrically isomorphic under the map* $x + P \to x' + P'$.

<u>Proof.</u> The map is an isomorphism since $rad(A) \subset P$ $(P \in \Pi(A))$. A straightforward computation shows that the mapping is an isometry ●

BA.2.7 *If* A *is a semisimple Banach algebra and* $e^2 = e \in A$, *then the closed subalgebra* eAe *is semisimple.*

<u>Proof.</u> $B = eAe$ is closed in A since e is idempotent. The map $P \to P \cap B$ is a homeomorphism of $\Pi(A) \setminus h(B)$ onto $\Pi(B)$ ((14) 26.14), so $rad(B) = rad(A) \cap B = (0)$ ●

The quasinilpotent characterisation of the radical in the next theorem is due to Zemánek (104). $\mathcal{Q}(A)$ denotes the set of quasinilpotent elements of A.

BA.2.8 *Let* A *be a unital Banach algebra, then*

(i) $rad(A)$ *contains any right or left ideal all of whose elements are quasinilpotent;*

(ii) $rad(A) = \{x \in A : x + \mathcal{Q}(A) \subset \mathcal{Q}(A)\}$;

(iii) $rad(A) = \{x \in A : x + Inv(A) \subset Inv(A)\}$.

<u>Proof.</u> (i) follows from (14) 24.18.

(ii), (iii) We show that

$$x + \mathcal{Q}(A) \subset \mathcal{Q}(A) \Rightarrow x \in rad(A) \Rightarrow x + Inv(A) \subset Inv(A) \Rightarrow x + \mathcal{Q}(A) \subset \mathcal{Q}(A).$$

Let $x + \mathcal{Q}(A) \subset \mathcal{Q}(A)$. Then $0 \in \mathcal{Q}(A) \Rightarrow x \in \mathcal{Q}(A)$. Suppose π is an irreducible representation of A on a linear space X and that $\pi(x) \neq 0$. Choose $\xi \in X$ such that $\pi(x)\xi \neq 0$, then, since $x \in \mathcal{Q}(A)$, ξ and $\pi(x)\xi$ are linearly independent. By the Jacobson density theorem ((14) 24.10) there exists $u \in A$ such that $\pi(u)\xi = 0$, $\pi(u)\pi(x)\xi = \xi$. Choose $0 \neq \lambda \in \rho(u)$ and put $v = \lambda - u$.

$$\begin{aligned}\pi(v^{-1}xv - x)\xi &= \pi(v^{-1})\pi(xv - vx)\xi,\\ &= \pi(v^{-1})\pi(ux - xu)\xi,\\ &= \pi(v^{-1})\xi = \lambda^{-1}\xi.\end{aligned}$$

Thus $\lambda^{-1} \in \sigma(v^{-1}xv - x)$. But $x \in \mathcal{Q}(A) \Rightarrow v^{-1}xv \in \mathcal{Q}(A)$, hence $x - v^{-1}xv \in \mathcal{Q}(A)$ by hypothesis, which is a contradiction. It follows that $x \in P$ $(P \in \Pi(A))$, hence $x \in \text{rad}(A)$.

Let $x \in \text{rad}(A)$. If $u \in \text{Inv}(A)$, $(u + x)^{-1} = u^{-1}(1 + xu^{-1})^{-1}$, hence $u + x \in \text{Inv}(A)$. Thus $x + \text{Inv}(A) \subset \text{Inv}(A)$.

Let $x + \text{Inv}(A) \subset \text{Inv}(A)$. Now $\lambda x + \text{Inv}(A) \subset \text{Inv}(A)$ $(\lambda \in \mathbb{C})$, and $q \in \mathcal{Q}(A) \iff 1 + \lambda q \in \text{Inv}(A)$ $(\lambda \in \mathbb{C})$. Thus $1 + \lambda(q + x) \in \text{Inv}(A)$ $(\lambda \in \mathbb{C})$, and it follows that $q + x \in \mathcal{Q}(A)$, therefore $x + \mathcal{Q}(A) \subset \mathcal{Q}(A)$ ●

BA.3 Minimal ideals and the socle

Let A be an algebra over $\mathbb{C}$, a *minimal right ideal* of A is a right ideal $J \neq (0)$, such that (0) and J are the only right ideals contained in J. A *minimal idempotent* is a non-zero idempotent e such that eAe is a division algebra. (If A is a Banach algebra $eAe = \mathbb{C}e$). The set of minimal idempotents in A is denoted by $\text{Min}(A)$. There are similar statements for left ideals.

BA.3.1 *If* A *is a semisimple algebra, then*

(i) R *is a minimal right ideal of* $A \iff R = eA$ *where* $e \in \text{Min}(A)$;

(ii) L *is a minimal left ideal of* $A \iff L = Af$ *where* $f \in \text{Min}(A)$;

(iii) $(1 - e)A, (A(1 - f))$ *is a maximal modular right (left) ideal of* A *if, and only if,* $e, f \in \text{Min}(A)$.

((14) 30.6, 30.11).

BA.3.2 (i) *Let* J *be a minimal right ideal of* A *and let* $u \in A$. *Then either* $uJ = (0)$, *or* uJ *is a minimal right ideal of* A.
(ii) *If* $x \in A$, $e \in \mathrm{Min}(A)$ *and* $xe \neq 0$ *then* xeA *is a minimal right ideal of* A.
((14) 30.7, (75) 2.1.8).

If A has minimal right ideals the smallest right ideal containing them all is called the *right socle* of A. If A has both minimal right and left ideals, and if the right and left socles of A are equal, we say that the *socle* of A exists and denote it by soc(A). Clearly the socle, if it exists, is a non-zero ideal of A. If A has no minimal left or right ideals we put $\mathrm{soc}(A) = (0)$.

BA.3.3 *Let* A *be a semisimple algebra with ideal* J. *Then*
(i) soc(A), soc(J) *exist;*
(ii) $\mathrm{Min}(J) = J \cap \mathrm{Min}(A)$;
(iii) $\mathrm{soc}(J) = J \cap \mathrm{soc}(A)$;
(iv) *if* A *is a Banach algebra and* $e, f \in \mathrm{Min}(A)$ *then* $\dim(eAf) \leq 1$.

Proof. (i) ((14) 30.10, 24.20).
(ii) straightforward.
(iii) follows from (ii) and BA.3.1.
(iv) ((14) 31.6).

BA.3.4 *Let* A *be a semisimple algebra,* $P \in \Pi(A)$, *and let* ϕ *denote the canonical quotient homomorphism* $\phi : A \to A/P$. *Then* A/P *is semisimple and* $\phi(\mathrm{soc}(A)) \subset \mathrm{soc}(\phi(A))$.

Proof. $\phi(\mathrm{Min}(A)) \subset \mathrm{Min}(\phi(A))$ and the result follows from BA.3.1 ●

The relationship between minimal idempotents and primitive ideals is important.

BA.3.5 *Let* A *be a semisimple algebra.*
(i) *If* $e \in \mathrm{Min}(A)$, *there exists a unique* $P_e \in \Pi(A)$ *such that* $e \notin P_e$.
(ii) *If* $e^2 = e \in \mathrm{soc}(A)$, *the set* $\{P \in \Pi(A) : e \notin P\}$ *is finite.*

Proof. (i) If $e \in \mathrm{Min}(A)$ then $A(1 - e)$ is a maximal modular left ideal (BA.3.1), therefore $P_e = \{x \in A : xA \subset A(1 - e)\}$ is a primitive ideal. Clearly $e \notin P_e$. If $Q \in \Pi(A)$ and $e \notin Q$, then $Q \cap Ae = (0)$, since Ae is a minimal left ideal. Thus $Qe = (O)$. Now if $q \in Q$, $qA \subset Q$ therefore $qAe = (O)$. It follows that $q \in P_e$, hence $Q \subset P_e$. On the other hand $P_e Ae = (O)$, and hence, by BA.2.1, either $Ae \subset Q$, or $P_e \subset Q$. But $Q \cap Ae = (O)$. Hence $P_e \subset Q$, giving $P_e = Q$.

(ii) If $e^2 = e \in soc(A)$, then $e = e_1 + \ldots + e_n$ where $e_i \in Min(A)$ $(1 \leq i \leq n)$. If $P \in \Pi(A)$ and $e \notin P$, then $e_i \notin P$ for some i. Therefore $P = P_{e_i}$ and the result follows ●

Information is also required on the set of accumulation points $\Pi^*(A)$ of $\Pi(A)$ in the hull-kernel topology.

BA.3.6 *If* A *is a semisimple algebra then* $\Pi^*(A) \subset h(soc(A))$.

Proof. Let $P \in \Pi(A)$ and $P \notin h(soc(A))$. Then there exists $x \in soc(A) \setminus P$. Thus $x = \sum_1^n a_i e_i$ where $a_i \in A$, $e_i \in Min(A)$ $(1 \leq i \leq n)$. Hence at least one e_i (e say) does not lie in P. So, by BA.3.5, $\Pi(A)$ is the disjoint union $\{P\} \cup h(\{e\})$. Now $h(\{e\})$ is closed in $\Pi(A)$, so $\{P\}$ is open, therefore $P \notin \Pi^*(A)$ ●

The Gelfand topology on the structure space of a commutative Banach algebra is, in general, stronger than the hull-kernel topology ((14) 23.4).

BA.3.7 *If* A *is a commutative Banach algebra then* $\Pi(A)$ *is discrete in the Gelfand topology* $\Longleftrightarrow$ $\Pi(A)$ *is discrete in the hull-kernel topology.*

Proof. Without loss of generality we may assume A to be semisimple. If $\Pi(A)$ is discrete in the hull-kernel topology, then it is clearly discrete in the Gelfand topology. Conversely, suppose that $\Pi(A)$ is discrete in the Gelfand topology. By the Šilov idempotent theorem ((13) 21.5), if $P \in \Pi(A)$, there exists $p = p^2 \in A$ such that $\hat{p}(P) = 1$, $\hat{p}(Q) = 0$ $(Q \in \Pi(A), Q \neq P)$ where $\hat{p}$ is the Gelfand transform of p. Then $p \in Min(A)$ and $p \notin P$, thus $\Pi(A)$ is the disjoint union $\{P\} \cup h(\{p\})$ by BA.3.5. Now $h(\{p\})$ is hull-kernel closed, so $\{P\}$ is hull-kernel open, hence $\Pi(A)$ is discrete in this topology ●

BA.3.8 *If* A *is a semisimple commutative Banach algebra such that* $\Pi(A)$ *is discrete then* $h(soc(A)) = \phi$.

Proof. From the above proof if $P \in \Pi(A)$, there exists $p \in Min(A)$ such that $p \notin P$, so $soc(A) \not\subset P$ ●

BA.3.9 *Let* A *be a unital semisimple Banach algebra such that* $\sigma(x)$ *is a singleton set for each* $x \in A$, *then* $A = \mathbb{C}1$.

Proof. Let $x \in A$ with $r(x) = 0$. Then we claim that $r(xy) = 0$ $(y \in A)$, for suppose there exists $y \in A$ such that $r(xy) > 0$, then by the hypothesis and (14) 5.3, $\sigma(xy) = \sigma(yx)$ is not zero. Thus $yx, xy \in \mathrm{Inv}(A)$, hence $x \in \mathrm{Inv}(A)$, which is a contradiction. Thus, by (14) 24.16, $r(x) = 0 \Longrightarrow x \in \mathrm{rad}(A)$, hence $x = 0$. Now, if $0 \neq y \in A$ then $\sigma(y) = \{\lambda\}$ where $\lambda \neq 0$ so $r(y - \lambda 1) = 0$ hence $y = \lambda 1$ ●

BA.3.10 *If* A *is a semisimple Banach algebra and if* p *is a non-zero idempotent which is not minimal, then there exists* $x \in pAp$ *such that* $\sigma_{pAp}(x)$ *contains two distinct points.*

Proof. pAp is a semisimple Banach algebra with unit p (BA.2.7), so if $\sigma_{pAp}(x)$ is a singleton set for each $x \in pAp$, by BA.3.9, $pAp = \mathbb{C}p$ and $p \in \mathrm{Min}(A)$ ●

BA.4 C*-algebras

A Banach algebra A is a C*-algebra if it possesses an involution * such that $||x^*x|| = ||x||^2$ $(x \in A)$. (The terminology B*-algebra is also used). The Gelfand-Naimark theorem states that every C*-algebra is isometrically *-isomorphic to a closed * subalgebra of $\mathcal{B}(H)$ for some Hilbert space H ((14) 38.10). The commutative version of the theorem, due to Gelfand, is as follows. Let A be a commutative C*-algebra then the space Ω of non-zero characters (multiplicative linear functionals) on A is locally compact in the weak * topology and A is isometrically-*-isomorphic to $C_o(\Omega)$; further, Ω is compact $\Longleftrightarrow$ A is unital ((14) 17.4, 17.5).

BA.4.1. *Let* A *be a C*-algebra, then*

(i) A *is semisimple;*

(ii) *if* I *is a closed ideal of* A, *then* $I = I^*$ *and* A/I *in the quotient norm is a C*-algebra;*

(iii) *if* ϕ *is a continuous *-homomorphism of* A *into a C*-algebra* B *then* $\phi(A)$ *is closed in* B.

((75) 4.1.19, 4.9.2, 4.8.5).

BA.4.2 *Let* A *be a unital* C^**-algebra and let* B *be a closed unital* $*$*-subalgebra of* A *then* $\sigma_B(x) = \sigma_A(x)$ $(x \in B)$.
((75) 4.8.2).

BA.4.3 *Let* A *be a* C^**-algebra.*

(i) *If* $f = f^2 \in A$, *there exists* $e = e^2 = e^* \in A$ *such that* $fe = e$ *and* $ef = f$.

(ii) *If* R *is a minimal right ideal of* A, *there exists* $e = e^* \in \text{Min}(A)$ *such that* $R = eA$.

(iii) *If* A *contains a right ideal* $R \subset \sum_1^n f_i A$ $(f_i \in \text{Min}(A),\ 1 \leq i \leq n)$ *there exists* $e = e^2 = e^* \in \text{soc}(A)$ *such that* $R = eA$.

<u>Proof.</u> (i) Using the Gelfand-Naimark representation this is equivalent to the elementary assertion that if a projection is contained in a C^*-algebra of operators on a Hilbert space then the C^*-algebra contains a self-adjoint projection with the same range ((84) 6.1).

(ii) If R is a minimal right ideal there exists $f^2 = f \in \text{Min}(A)$ such that $R = fA$. By (i) find $e = e^2 = e^* \in A$ such that $fe = e$, $ef = f$. Then $R = fA = efA \subset eA = feA \subset fA$. Thus $R = eA$, hence $e \in \text{Min}(A)$.

(iii) Similar argument ●

It is elementary to check the uniqueness of the self-adjoint idempotents in BA.4.3. Since a C^*-algebra is semisimple its socle exists.

BA.4.4 *Let* A *be a* C^**-algebra, then*

(i) $\text{soc}(A) = (\text{soc}(A))^*$;

(ii) $x \in \text{soc}(A) \iff x^*x \in \text{soc}(A)$;

(iii) $x \in \overline{\text{soc}(A)} \iff x^*x \in \overline{\text{soc}(A)}$.

<u>Proof.</u> (i) If $x \in \text{soc}(A)$, then $x \in R \subset \sum_1^n f_i A$ where R is a right ideal of A and each $f_i \in \text{Min}(A)$. By BA.4.3, $R = eA$ where $e = e^2 = e^* \in \text{soc}(A)$. So $x = ex$, hence $x^* = x^*e \in Ae \subset \text{soc}(A)$.

(ii) $\Rightarrow$ is clear. Let $x \in A$ and suppose that $x^*x \in \text{soc}(A)$. Then there exists $e = e^2 = e^* \in \text{soc}(A)$ such that $x^*x \in Ae$ (BA.4.3). Thus $x^*x(1 - e) = 0$, and

$$||x - xe||^2 = ||x(1 - e)||^2 = ||(1 - e)x^*x(1 - e)|| = 0,$$

so $x = xe \in soc(A)$.

(iii) Let I be a closed ideal of the C*-algebra A. Then $I = I^*$ and A/I is a C*-algebra (BA.4.1), hence

$$||x^*x + I|| = ||(x^* + I)(x + I)|| = ||x + I||^2.$$

So $x^*x \in I \Leftrightarrow x \in I$ ●

Finally we need a result on the spectrum of an operator matrix. $\text{int}(\Omega)$ denotes the interior of the set Ω, and $U, V \in \mathcal{B}(H)$.

BA.4.5 *If* $\text{int}(\sigma(U) \cap \sigma(V)) = \phi$ *and*

$$T = \begin{bmatrix} U & * \\ 0 & V \end{bmatrix}$$

then $\sigma(T) = \sigma(U) \cup \sigma(V)$.

This follows immediately from the following lemma.

BA.4.6 $(\sigma(U) \cup \sigma(V)) \setminus \text{int}(\sigma(U) \cap \sigma(V)) \subset \sigma(T) \subset \sigma(U) \cup \sigma(V)$.

<u>Proof.</u> Elementary matrix computation shows that

$$(\sigma(U) \cup \sigma(V)) \setminus (\sigma(U) \cap \sigma(V)) \subset \sigma(T) \subset \sigma(U) \cup \sigma(V).$$

Now choose $\lambda \in \partial(\sigma(U) \cap \sigma(V))$ then $\lambda \in \partial\sigma(U) \cup \partial\sigma(V)$ so either $\lambda - U$ or $\lambda - V$ is a two-sided topological divisor of zero in the Banach algebra of all bounded linear operators. In the first case there exist A_n with $||A_n|| = 1$ for each n, and $(\lambda - U)A_n \to 0$.

Then $$(\lambda - T)\begin{bmatrix} A_n & 0 \\ 0 & 0 \end{bmatrix} = \begin{bmatrix} (\lambda - U)A_n & 0 \\ 0 & 0 \end{bmatrix} \to 0,$$

So $\lambda \in \sigma(T)$. In the other case, there exist B_n with $||B_n|| = 1$ for each n, and $B_n(\lambda - V) \to 0$, hence

$$\begin{bmatrix} 0 & 0 \\ 0 & B_n \end{bmatrix} (\lambda - T) = \begin{bmatrix} 0 & 0 \\ 0 & B_n(\lambda - V) \end{bmatrix} \to 0,$$

again $\lambda \in \sigma(T)$. Thus $\partial(\sigma(U) \cap \sigma(V)) \subset \sigma(T)$ ●

It is easy to see that the result of BA.4.5 fails if we drop the condition that $\mathrm{int}(\sigma(U) \cap \sigma(V)) = \phi$. Take $H = \ell^2 \oplus \ell^2$ and if T is the bilateral shift on H,

$$T = \begin{bmatrix} U & * \\ 0 & V \end{bmatrix},$$

where U and V are the forward and backward shifts on ℓ^2, $\sigma(U)$ and $\sigma(V)$ are the unit disk, while $\sigma(T)$ is the unit circle.

Bibliography

1. Ambrose, W., Structure theorems for a special class of Banach algebras, T.A.M.S. 137, 1969.

2. Akemann, C.A., Pedersen, G., Ideal perturbations of elements in C^*-algebras, Math. Scand. 41, 1977.

3. Akemann, C.A., Wright, S., Compact actions on C^*-algebras, Glasgow Math. J. 21, 1980.

4. Alexander, J.C., Compact Banach algebras, P.L.M.S. (3) 18, 1968.

5. Atkinson, F.V., The normal solvability of linear equations in normed spaces, Mat. Sbornik 28 (70), 1951.

6. Aupetit, B., Propriétés Spectrales des Algèbres de Banach, Lecture Notes in Mathematics 735 (Springer) Berlin, 1979.

7. Barnes, B.A., A generalised Fredholm theory for certain maps in the regular representation of an algebra, Can. J. Math. 20, 1968.

8. Barnes, B.A., The Fredholm elements of a ring, Can. J. Math. 21, 1969.

9. Barnes, B.A., Examples of modular annihilator algebras, Rocky Mountain J. Math. 1, 1971.

10. Berglund, M.F., Ideal C^*-algebras, Duke Math. J. 40, (1973).

11. Bonsall, F.F., Compact operators from an algebraic standpoint, Glasgow Math. J. 8, 1967.

12. Bonsall, F.F., Operators that act compactly on an algebra of operators, B.L.M.S. 1, 1969.

13. Bonsall, F.F., Compact Linear Operators, Yale University Lecture Notes, 1967.

14. Bonsall, F.F., Duncan, J., Complete Normed Algebras, (Springer) Berlin, 1973.

15. Bonsall, F.F., Goldie, A.W., Annihilator algebras, P.L.M.S. (3) 4, 1954.

16. Browder, F.E., On the spectral theory of elliptic differential operators I, Math. Ann. 142, 1961.

17. Buoni, J.J., Harte, R.E., Wickstead, A., Upper and lower Fredholm spectra, P.A.M.S. 66, 1977.

18. Breuer, M., Fredholm theories in von-Neumann algebras I, Math. Ann. 178, 1968.

19. Breuer, M., Fredholm theories in von-Neumann algebras II, Math. Ann. 180, 1969.

20. Calkin, J.W., Two-sided ideals and congruences in the ring of bounded operators in Hilbert space, Ann. Math. (2) 42, 1941.

21. Caradus, S.R., Operators of Riesz type, Pacific J. Math. 18, 1966.

22. Caradus, S.R., Perturbation theory for generalised Fredholm operators, Pac. J. Math. 52, 1974.

23. Caradus, S.R., Perturbation theory for generalised Fredholm operators II, P.A.M.S. 62, 1977.

24. Caradus, S.R., Operator Theory of the Pseudo-Inverse, Queen's Papers in Pure and Appl. Math. 38, Kingston, 1977.

25. Caradus, S.R., Pfaffenberger, W., Yood, B., Calkin Algebras and Algebras of Operators on Banach Spaces, (Dekker) New York, 1974.

26. Chui, C.K., Smith, P.W., Ward, J.D., A note on Riesz operators, P.A.M.S. 60, 1976.

27. Dixmier, J., C^*-Algebras, (North-Holland) Amsterdam, 1977.

28. Douglas, R.G., On majorization, factorization and range-inclusion of operators in Hilbert space, P.A.M.S. 17, 1966.

29. Dowson, H.R., Spectral Theory of Linear Operators, (Academic Press) London, 1978.

30. Dunford, N., Schwartz, J.T., Linear Operators, Part I, (Wiley-Interscience) New York, 1958.

31. Erdos, J.A., On certain elements of C^*-algebras, Illinois J. Math. 15, 1971.

32. Erdos, J.A., Giotopoulos, S., Lambrou, M.S., Rank one elements of Banach algebras, Mathematika, 24, 1977.

33. Fredholm, I., Sur une classe d'équations fonctionelles, Acta. Math. 27, 1903.

34. Freundlich, M., Completely continuous elements of a normed ring, Duke Math. J. 16, 1949.

35. Gillespie, T.A., West, T.T., A characterisation and two examples of Riesz operators, Glasgow Math. J. 9, 1968.

36. Gohberg, I.C., On linear equations in normed spaces, Dokl. Akad. Nauk. S.S.S.R. 76, 1951.

37. Gohberg, I.C., On linear equations depending analytically on a parameter, Dokl. Akad. Nauk. S.S.S.R. 78, 1951.

38. Gramsch, B., Lay, D.C., Spectral mapping theorems for essential spectra, Math. Ann. 192, 1971.

39. Halmos, P.R., Quasitriangular operators, Acta Sci. Math.(Szeged) 29, 1968.

40. de la Harpe, P., Initiation à l'algèbre de Calkin, Lecture Notes in Mathematics 725 (Springer) Berlin 1978.

41. Harte, R.E., Wickstead, A., Upper and lower Fredholm spectra II, Math. Zeit. 154, 1977.

42. Heuser, H., Zur Eigenwerttheorie einer Klasse Rieszscher Operatoren, Archiv. Math. 14, 1963.

43. Heuser, H., Atkinson-und Fredholm-operatoren, Lecture Notes, Karlsruhe, 1969.

44. Heuser, H., Funktionalanalysis, (Teubner) Stuttgart, 1975.

45. Hewitt, E., Ross, K.A., Abstract Harmonic Analysis Vol. II, (Springer) Berlin, 1970.

46. Hilbert, D., Grundzüge einer allgemeinen Theorie der linearen Integralgleichungen IV, Nachr. Akad, Wiss. Göttingen Math.-Phys. Kl. 1906.

47. Huruya, T., A spectral characterisation of a class of C*-algebras, Sci. Rep. Niigata Univ. Ser. A 15 (1978).

48. Jacobson, N., Structure of Rings, (A.M.S.) Providence, 1956.

49. Kaashoek, M.A., Smyth, M.R.F., On operators T such that f(T) is a Riesz or meromorphic, P.R.I.A. 72(A), 1972.

50. Kaplansky, I., Dual rings, Ann. Math. (2) 49, 1948.

51. Kaplansky, I., Normed algebras, Duke Math. J. 16, 1949.

52. Kato, T., Perturbation theory for nullity, deficiency, and other quantities of linear operators, J. d'Analyse Math. 6, 1958.

53. Kleinecke, D.C., Almost-finite, compact and inessential operators, P.A.M.S. 14, 1963.

54. Kroh, H., Saturierte Algebren, Math. Ann. 211, 1974.

55. Laffey, T.J., West, T.T., Fredholm commutators, P.R.I.A. 82(A), 1982.

56. Lay, D.C., Characterisations of the essential spectrum of F.E. Browder, B.A.M.S. 74, 1968.

57. Lebow, A., Schechter, M., Semigroups of operators and measures of non-compactness, J. Functional Anal. 7, 1971.

58. Legg, D., A note on Riesz elements in C*-algebras, Internat. J. Math. and Math. Sci. 1, 1978.

59. Mizori-Oblak, P., Fredholm elements in Banach algebras, Glasnik Mat. 10, 1975.

60. Murphy, G.J., West, T.T., Spectral radius formulae, P. Edinburgh Math. Soc. 22, 1979.

61. Murphy, G.J., West, T.T., Decomposition algebras of Riesz operators, Glasgow Math. J. 21, 1980.

62. Murphy, G.J., West, T.T., Decomposition of index-zero Fredholm operators, P.R.I.A. 81(A), 1981.

63. Murphy, I.S., Non-compact operators that act compactly on their centralisers, B.L.M.S. 2, 1970.

64. Nashed, M.Z., Generalised Inverses and Applications, (Academic Press) New York, 1976.

65. Ogasawara, T., Finite dimensionality of certain Banach algebras, J. Sci. Hiroshima Univ. 17(A), 1954.

66. Ogasawara, T., Yoshinaga, K., Weakly completely continuous Banach *-algebras, J. Sci. Hiroshima Univ. 18(A), 1954.

67. Olsen, C.L., A structure theorem for polynomially compact operators, Amer. J. Math. 93, 1971.

68. Olsen, C.L., A concrete representation of index theory in von-Neumann algebras, Lecture Notes in Mathematics 693 (Springer) Berlin, 1977.

69. Pearlman, L.D., Riesz points of the spectrum of an element in a semi-simple Banach algebra, T.A.M.S. 193, 1974.

70. Pedersen, G.K., Spectral formulas in quotient C*-algebras, Math. Zeit. 148, 1976.

71. Przeworska-Rolewicz, D., Rolewicz, S., Equations in Linear Spaces, (P.W.N.) Warsaw, 1968.

72. Puhl, J., The trace of finite and nuclear elements in Banach algebras, Czech. Math. J. 28, 1978.

73. Radjavi, H., Laurie, C., On the West decomposition of Riesz operators, B.L.M.S. 12, 1980.

74. Riesz, F., Über lineare Funktionalgleichungen, Acta. Math. 41 (1918).

75. Rickart, C.E., General Theory of Banach Algebras, (Van Nostrand) Princeton, 1960.

76. Ruston, A.F., Operators with a Fredholm theory, J.L.M.S., 29, 1954.

77. Salinas, N., Operators with essentially disconnected spectrum, Acta. Sci. Math. 33, 1972.

78. Schechter, M., Basic theory of Fredholm operators, Ann. Scuola Normale Sup. di Pisa, Classe di Sci. 21, 1967.

79. Schechter, M., On the essential spectrum of an arbitrary operator, J. Math. Anal. Appl. 13, 1966.

80. Schechter, M., Principles of Functional Analysis, (Academic Press) New York, 1971.

81. Smyth, M.R.F., A note on the action of an operator on its centraliser, P.R.I.A. 74(A), 1974.

82. Smyth, M.R.F., Ideals of algebraic elements of a Banach algebra, Trinity College, Dublin, School of Mathematics Preprints 11, 1974.

83. Smyth, M.R.F., Fredholm theory in Banach algebras, Trinity College, Dublin, School of Mathematics Preprints 8, 1975.

84. Smyth, M.R.F., Riesz theory in Banach algebras, Math. Zeit. 145, 1975.

85. Smyth, M.R.F., Riesz algebras, P.R.I.A. 76(A), 1975.

86. Smyth, M.R.F., On problems of Olubummo and Alexander, P.R.I.A. 80(A), 1980.

87. Smyth, M.R.F., West, T.T., The spectral radius formula in quotient algebras, Math. Zeit. 145, 1975.

88. Stampfli, J.G., Compact perturbations, normal eigenvalues and a problem of Salinas, J.L.M.S. (2) 9, 1974.

89. Tomiuk, B.J., Wong, P.K., Weakly semi-completely continuous A*-algebras, Illinois J. Math 16, 1972.

90. Treese, G.W., Kelly, E.P., Generalised Fredholm operators and the boundary of the maximal group, P.A.M.S. 67, 1977.

91. Vala, K., On compact sets of compact operators, Ann. Acad. Sci. Fenn. AI, 1964.

92. Vala, K., Sur les éléments compacts d'une algèbre normée, Ann. Acad. Sci. Fenn. AI, 1968.

93. Veselić, K., On essential spectra in Banach algebras, Glasnik Mat. 10 (3), 1975.

94. West, T.T., Riesz operators in Banach spaces, P.L.M.S. (3) 16, 1966.

95. West, T.T., The decomposition of Riesz operators, P.L.M.S. (3) 16, 1966.

96. Wong, P.K., Modular annihilator A*-algebras, Pac. J. Math. 37, 1971.

97. Yang, K.-W., The generalised Fredholm operators, T.A.M.S. 216, 1976.

98. Yang, K.-W., Operators invertible modulo the weakly compact operators, Pacific J. Math. 71, 1977.

99. Ylinen, K., Compact and finite dimensional elements of normed algebras, Ann. Acad. Sci. Fenn. AI, 1968.

100. Ylinen, K., A note on the compact elements of C*-algebras, P.A.M.S. 35, 1972.

101. Ylinen, K., Weakly completely continuous elements of C*-algebras, P.A.M.S. 52, 1975.

102. Yood, B., Difference algebras of linear transformations on a Banach space, Pacific J. Math. 4, 1954.

103. Yood, B., Homomorphisms on normed algebras, Pacific J. Math. 8, 1958.

104. Zemánek, J., A note on the radical of a Banach algebra, Manuscripta Math. 20, 1977.

SUPPLEMENTARY BIBLIOGRAPHY

105. Kirchberg, E., Banach algebras whose elements have at most countable spectra, (preprint) Akad. Wiss. DDR, Zentralinstitut Math. Mech. 1979.

106. Harte, R.E., Fredholm theory relative to a Banach algebra homomorphism, Math. Zeit. 179, 1982.

107. Zemánek, J., The essential spectral radius and the Riesz part of the spectrum, Functions, Series, Operators (Proc. Internat. Conf., Budapest, 1980) Colloq. Math. János Boÿlài (to appear).

108. Zemánek, J., Generalisations of the spectral radius formula, P.R.I.A. 81(A), 1981.

109. Kraljević, H., Veselić, K., On algebraic and spectrally finite Banach algebras, Glasnik Mat. 11 (31), 1976.

110. Kraljević, H., Suljagić, S., Veselić, K., Index in semisimple Banach algebras, preprint Dept. of Math. Univ. of Zagreb 1980.

Index

Notation